Experimental and Theoretical Study on Thermal Consolidation Behavior of Saturated Clay

Guangchang Yang · Bing Bai

Experimental and Theoretical Study on Thermal Consolidation Behavior of Saturated Clay

Springer

Guangchang Yang
School of future cities
University of Science and Technology Beijing
Beijing, China

Bing Bai
School of Civil Engineering
Beijing Jiaotong University
Beijing, China

ISBN 978-981-96-6908-0 ISBN 978-981-96-6909-7 (eBook)
https://doi.org/10.1007/978-981-96-6909-7

This Springer imprint is published by the registered company Springer Nature Singapore Pte Ltd.
The registered company address is: 152 Beach Road, #21-01/04 Gateway East, Singapore 189721, Singapore

Contents

Chapter 1
Introduction

1.1 Research Background

With the vigorous development of infrastructure in society, more and more geotechnical engineering problems have emerged in front of us, such as water conservancy and hydropower projects, nuclear waste storage, energy foundations, oil and gas extraction, deep foundation pit excavation, temperature effects of high-voltage cables on surrounding soil, as well as deformation and stability problems caused by environmental temperature changes in road, slope, municipal, and other engineering constructions [1–3]. These engineering problems mostly involve the coupling effects between multiple fields such as stress field (M), seepage field (H), temperature field (T), and chemical field (C) of rock and soil. Among them, the temperature field plays an important role in the coupling effect with other fields, but it is often overlooked by people. For example, temperature can lead to a decrease in the mechanical properties of soil, intensify the vaporization of pore water, cause changes in the viscosity of pore water, and result in changes in seepage velocity. These changes in turn affect the stress changes in soil, leading to changes in the mechanical properties of geotechnical materials. Obviously, temperature changes can have a significant impact on the engineering properties of soil, leading to a series of engineering problems such as soil deformation, stability, and reduced bearing capacity.

With the increasing global attention to environmental issues, new energy sources such as geothermal energy resources are receiving more and more attention. In the development of shallow geothermal energy, structures such as pile foundations or underground continuous walls are often used as thermal storage carriers for underground spaces, also known as energy foundations [4]. Energy based is a rapidly developing geothermal exchange system that not only meets the technical requirements of geotechnical engineering foundations, but also serves as an energy-saving system for heating and cooling buildings, thereby reducing energy consumption and carbon dioxide emissions [5, 6]. However, this energy-based heating and cooling can cause changes in the engineering properties of the surrounding soil, but this is

G. Yang and B. Bai, *Experimental and Theoretical Study on Thermal Consolidation Behavior of Saturated Clay*, https://doi.org/10.1007/978-981-96-6909-7_1

often overlooked in the analysis and design of engineering construction, which will have a certain impact on the working performance of the foundation. In addition, in underground engineering such as subway tunnels and deep excavation pits, precipitation measures are often required, which will cause changes in the soil moisture content, resulting in changes in the internal stress field and seepage field of the soil, ultimately leading to deformation of the soil, adverse effects on the surrounding environment, and even causing engineering accidents. High-voltage cables, heating pipelines, and other equipment buried underground will generate significant heat during long-term operation, forming a heat source carrier and changing the distribution of soil temperature field [7]. In underground waste treatment projects, such as landfill and nuclear waste disposal, soil becomes an important barrier to prevent the migration of pollutants, and waste treatment also generates a large amount of heat, which will change the temperature of the surrounding soil. Therefore, this process involves the coupling of heat, water, force, and chemical fields [8–10]. The study of thermal solidification of soil media is a product of the intersection and permeation of disciplines such as geotechnical mechanics, seepage mechanics, heat transfer, and engineering technology. The research on this problem initially started in the water conservancy and hydropower industries, and now it has spread to fields such as geothermal energy, natural gas hydrate extraction, and nuclear waste storage. Geotechnical media are products naturally generated over a long period of geological time and are porous media composed of different mineral combinations. It is influenced by various factors in nature, almost all of which interact in the form of fields and constantly change with the changes in time and space. The mechanical, physical, and chemical properties of rock and soil comprehensively reflect the deformation, failure, and interaction between various components under external stress and seepage fields, temperature fields, chemical fields, and other multiple fields, namely coupling effects. This coupling effect is often unidirectional or bidirectional and produces irreversible effects, such as fluid velocity and pressure, stress and deformation, temperature gradients, and processes between fluid solid chemical reactions. The multi-field coupling problem of thermal solidification of soil mainly focuses on the rock and soil mass, its occurrence environment, and engineering effects. By combining experimental and theoretical research, qualitative and quantitative analysis methods are used to study the multi-field coupling mechanism and coupling model, the numerical calculation method of the multi-field coupling model, and the deformation and failure mechanism of rock and soil mass under multi-field coupling. In terms of the application of multi-field coupling models in geotechnical media, the main focus is on the application research of hydraulic coupling models, such as in the fields of earth rock dam engineering and slope engineering. Other multi-field coupling models have gradually been applied in many fields such as geothermal development, waste treatment, and energy infrastructure engineering. However, the guiding role of multi-field coupling analysis theory and methods in engineering construction is relatively limited, and there are very few practical applications. Currently, most research focuses on the coupling effect of stress field and seepage field, and there is relatively little research on the coupling of thermal and hydraulic fields, which is an urgent need to be filled in the multi-field coupling theory.

From the above, it can be seen that with the development of society, the environment in which engineering construction is carried out is becoming increasingly complex, and the difficulty of engineering is increasing. Geotechnical mechanics and engineering are facing severe challenges. Therefore, there is an urgent need for the support of multi-field coupling theory and analysis technology for rock and soil mass, in order to improve the properties of rock and soil engineering, save construction investment, improve resource extraction efficiency, and enhance disaster prevention and mitigation capabilities.

1.2 Research Overview

1.2.1 Experimental Study on the Influence of Temperature on the Mechanical Properties of Soil

(1) The Effect of Temperature on Thermal Strain

Baldi et al. [11] conducted a systematic study on the volume change of heated saturated soil using a temperature-controlled triaxial test apparatus. By heating soil samples with three different overconsolidation ratios (OCR), it was found that different soil samples have similar thermal responses, and the evolution of thermal strain in different OCR soil samples is also different, as shown in Fig. 1.1. When OCR = 1, heating only causes thermal shrinkage of the soil sample; when OCR = 2.5, thermal expansion occurs at the initial temperature rise, and when the temperature rises to about 60 °C, it changes from thermal expansion to thermal contraction; When OCR = 12.5, heating causes significant thermal expansion, and as the temperature increases, the thermal expansion rate slightly decreases. Many scholars [12–17] have also discovered similar phenomena through thermal solidification tests on saturated or unsaturated soils. When the soil overconsolidation ratio is small, heating up will cause thermal shrinkage, and as the overconsolidation ratio increases, the amount of thermal shrinkage decreases. When the OCR exceeds a certain threshold, heating will cause thermal expansion. In addition, the study also found that the thermal shrinkage characteristics of soil are irreversible, while the thermal expansion characteristics are reversible behavior. Ghembazaa et al. [18] conducted experimental research on the thermal mechanical coupling effect of sandy clay under isotropic paths. The strain of the thermal solid at different temperatures showed that the compressive modulus tended to decrease at higher temperatures.

Tang et al. [19] studied the thermal strain characteristics of an unsaturated expansive clay using an isotropic experimental apparatus with controllable suction and temperature. A series of experimental studies were conducted on the changes in body strain during the heating cooling process under constant suction and net stress. Research has shown that under temperature loads, the volumetric strain of soil samples is not only affected by pressure, but also by suction. Under high-pressure

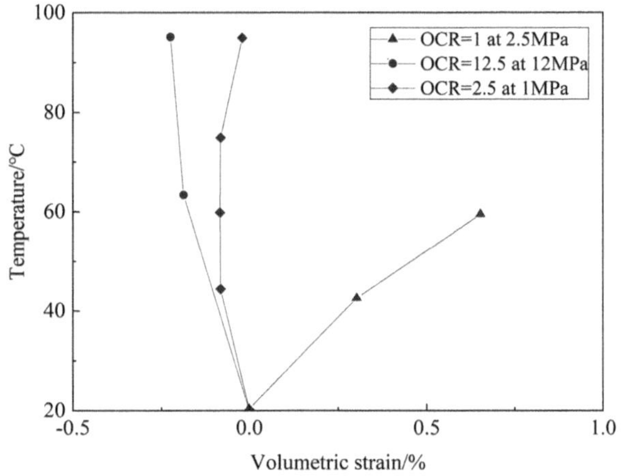

Fig. 1.1 Volume change of saturated Pontida clay samples with different OCRs during heating–cooling process [11]

and low-suction conditions, heating causes thermal shrinkage, while conversely, it causes thermal expansion, as shown in Fig. 1.2. This phenomenon can also be seen as related to the size of the soil sample OCR. Uchaipichat and Khalili [20] conducted a series of suction controlled heating tests on unsaturated compressed clay and found a similar phenomenon. They also pointed out that under the same pressure, the higher the suction, the more the soil sample tends to shrink due to heating. It seems that the experimental results obtained by the two are opposite, mainly due to different initial conditions, which result in different OCR of the soil sample during heating. Although the OCR of unsaturated soil is relatively easier to determine than that of saturated soil, it shows the phenomenon of thermal shrinkage when OCR is small and thermal expansion when OCR is large. Obviously, more experiments and theoretical research are needed to fully understand the thermal response of unsaturated soils during heating and cooling processes.

(2) Temperature-Induced Changes in Pore Water Pressure

Abuel-Naga et al. [17] conducted a series of undrained heating cooling experiments on reshaped saturated clay with different OCRs and found that undrained heating of saturated samples induced positive pore pressure, followed by cooling to reduce pore pressure and negative pore pressure, as shown in Fig. 1.3. The difference in thermal expansion coefficient between soil particles and pore water is the direct cause of positive and negative pore water pressure [21]. The thermal expansion coefficient of pore water is much larger than that of soil particles, about 17 times. When heated, the expansion deformation of pore water is much greater than that of soil particles, and due to the limitation of non-drainage, positive pore pressure will be generated. Similarly, during cooling, the shrinkage deformation of pore water is much greater than that of soil particles, and negative pore pressure will be generated due to the limitation of undrained conditions [16, 21–23]. From Fig. 1.3, it can also be seen that after a heating cooling cycle, there is an irreversible change in pore pressure. For normally consolidated soil (OCR = 1), this change is very small and can be

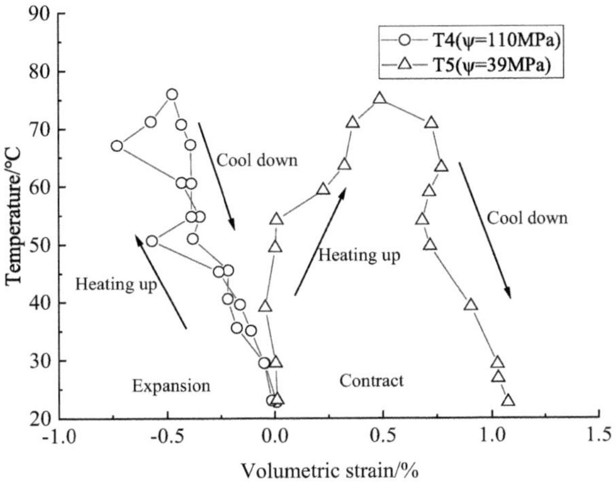

Fig. 1.2 Effect of suction on thermal volumetric strain of unsaturated soil [19]

ignored. As the OCR value increases, the amplitude of irreversible changes in pore pressure also increases. This change is related to the hysteresis phenomenon caused by loading and unloading. When the temperature rises, the pore pressure increases and the effective stress decreases, which is equivalent to unloading, while the opposite is loading.

(3) The Influence of Temperature on the Normal Compression Consolidation Curve (NCL)

Campanella and Mitchell [22] conducted a series of isotropic compression tests on saturated remolded clay in the range of 25–50 °C. When the temperature rises, the normal compression consolidation curve moves downward in the semi-logarithmic coordinate (v–ln p') plane of specific volume effective stress. The compression curves

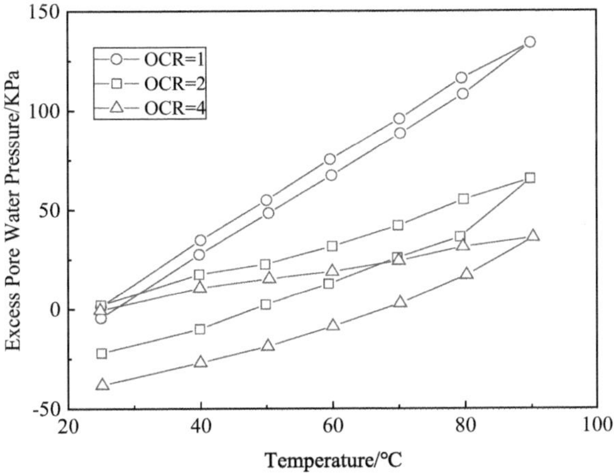

Fig. 1.3 Pore water pressure change caused by undrained heating–cooling process [17]

at different temperatures are relatively parallel, that is, the slope λ of the consolidation curve is not affected by temperature, as shown in Fig. 1.4. Other scholars have also reached the same conclusion in thermal effect experiments on saturated soils [12, 13] and unsaturated soils [19, 20, 24]. It should be emphasized that for the vast majority of soils, the slope λ of the normal compression curve is not affected by temperature, while [24] found that the value of the slope λ of the compression curve increased by about 15% as the temperature increased from 22 to 80 °C. Further experimental observation is needed for research in this area.

(4) The Influence of Temperature on Critical State Parameters

Some scholars [12, 13, 21, 25] have studied the effect of temperature on the critical state curve (CSL) using a temperature-controlled triaxial apparatus. The study found that, similar to the normal consolidation compression curve, as the temperature increases, the critical state curve also moves downward in the v–ln p' plane, but the magnitude of the two movements is not exactly the same. Some data shows that the movement amplitude of the critical state line is smaller than that of the normal consolidation compression curve [12, 13] and the distance between the two decreases with increasing temperature. On the contrary, there is also data showing that the distance between the two increases with increasing temperature [26, 27]. Figure 1.5 shows the effect of temperature on the critical state stress ratio M (q/p') [28], and it can be seen that different types of soils are also affected to varying degrees. For Boom clay and kaolin, the temperature increased by about 10% from 20 to 90 °C. For other types of soil, the value of M is basically not affected by temperature. There are few literature studies on the effect of temperature on the critical state of unsaturated soil. Based on existing literature, it is shown that the effect of temperature on the critical state of unsaturated soil is similar to that of saturated soil [20].

(5) The Influence of Temperature on the Yield Surface

Cekerevac and Laloui [13] found through temperature-controlled triaxial tests on saturated kaolin that temperature has a certain influence on the shape of the yield surface in the p'–q plane. As shown in Fig. 1.6, with the increase of temperature, the yield stress ratio M decreases. Similar phenomena have also been observed by other scholars [17, 25, 29] in drained and undrained triaxial shear tests. Some scholars [30, 31] believe that the effect of temperature on the yield surface is due to changes in the internal structure of the soil caused by heating. Eriksson [32] studied the effect of temperature on the initial consolidation pressure of unsaturated silty clay through consolidation compression tests, as shown in Fig. 1.7. It can be seen that the preconsolidation pressure decreases with increasing temperature, indicating that the elastic range of the soil decreases during the heating process, which is known as thermal softening. Uchaipichat and Khalili [20] found that the greater the suction, the more pronounced this effect becomes. Similarly, for saturated soil, heating also causes thermal softening of the soil [19, 20, 24].

For highly consolidated soil, the stress state remains within the yield surface during the heating process, and the thermal softening phenomenon will reduce the OCR of the soil. Due to the reduction of OCR, the soil will appear softer at high

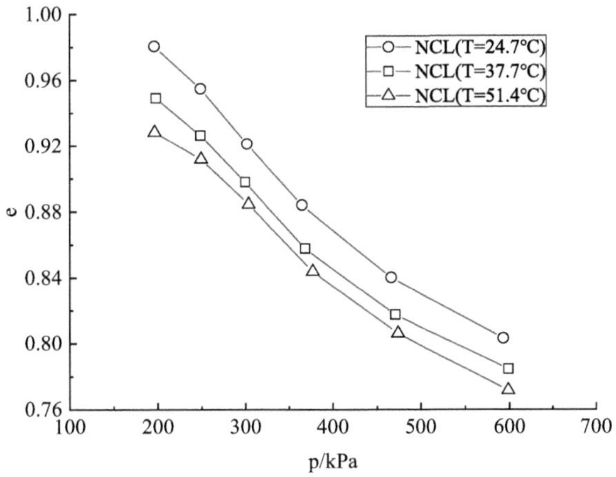

Fig. 1.4 Effect of temperature on normal consolidation curve of saturated remolded clay [22]

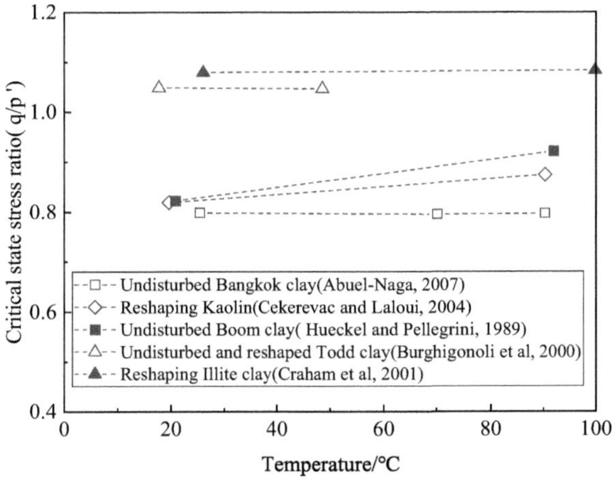

Fig. 1.5 Effect of temperature on stress ratio *M* under critical state [28]

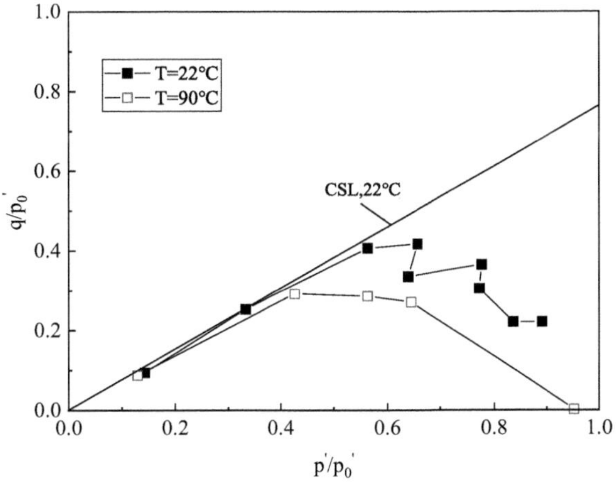

Fig. 1.6 Effect of temperature on yield curve on p'–q plane [13]

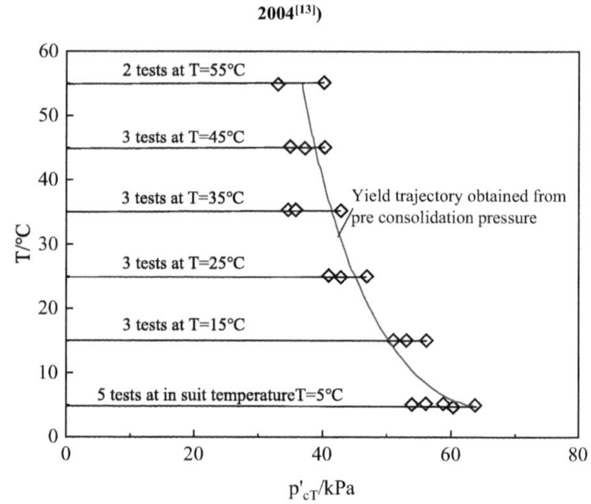

Fig. 1.7 Effect of temperature on preconsolidation pressure [32]

temperatures. Hueckel and Baldi [29] confirmed this phenomenon through compression tests on overconsolidated saturated clay. Although the effect of heating on the critical state friction angle can be almost negligible, as the temperature increases, the expansion angle and peak friction angle both decrease, which is also a reason for thermal softening. For normally consolidated soil, the stress state is always on the yield surface. When heated, volume shrinkage occurs, causing the yield surface to move outward, that is, strain hardening occurs. This strain hardening is offset by the thermal softening caused by heating, so normally consolidated soils may also exhibit hardening phenomena at high temperatures [33].

(6) The Influence of Temperature on Shear Characteristics

The influence of temperature on the drained and undrained shear strength of soil is quite complex, mainly depending on the type of soil and the initial conditions in which it is located. Some scholars [13, 34, 35] conducted experimental research on the effect of temperature on shear strength and provided corresponding experimental evolution laws. From these basic laws, it can be seen that for soil samples with different overconsolidation ratios, as the temperature increases, the shear strength (drained and undrained) increases to a certain extent, and the shear softening characteristics become more and more obvious.

Cekerevac and Laloui [33] studied the effect of temperature on the cyclic shear characteristics of normally consolidated saturated kaolin under undrained conditions. Two sets of cyclic shear tests were conducted at both 22 and 90 °C. The study found that the axial strain and excess pore water pressure caused by a single cyclic shear at 90 °C were smaller than those obtained at 22 °C. The axial strain at the final failure of the two sets of experiments is basically the same, but more cycles are required at high temperatures, which is related to the thermal deformation response caused by heating. As mentioned earlier, the overconsolidation ratio and suction have a significant impact on the thermal response of soil. This experiment is only for the case of normally consolidated saturated soil, and there is almost no literature on the cyclic shear temperature effect of unsaturated soil and overconsolidated soil. More research should be conducted in this area.

(7) The Effect of Temperature on Permeation Thermal Properties

The influence of temperature on permeability coefficient is not only related to the dynamic viscosity coefficient of pore fluid, but also to the internal structure of soil, mineral composition, and ion concentration and type in the fluid. The measurement of permeability coefficient is generally obtained based on the consolidation compression curve, also known as indirect testing. However, there are few direct tests for measuring soil permeability at different temperatures, including hard wall permeability tests and flexible wall permeability tests. Burghingoli et al. [36] and Towhata et al. [37] analyzed the advantages and disadvantages of using indirect testing methods for permeability and compared them with direct methods. Delage et al. [14] studied the variation of permeability with temperature by conducting constant head measurements at different temperatures. Indirect analysis based on consolidation parameter estimation indicates that indirect methods for estimating permeability from consolidation tests should be carefully considered. More scholars have adopted the hard wall permeability test method for permeability research, such as Morin et al. [38], who conducted constant head permeability tests on marine soils under different temperature conditions using rigid consolidation rings. Finn [39] believed that temperature has a relatively small effect on the viscosity of pore water, while Habibagahi [40] found through experimental research that heating can reduce the viscosity of water, leading to an increase in permeability coefficient. This conclusion was later confirmed by Rujivipat [41]. The experimental study by Ye et al. [42] showed that there is a certain difference between the estimated value and the measured value based on the permeability coefficient theory of pore water viscosity changes. As the temperature increases, the difference between the two becomes more significant.

The permeability coefficient of soil medium is generally expressed by the following equation:

$$k = K\frac{\rho g}{\mu} = K\frac{g}{\eta}. \tag{1.1}$$

Among them, μ is the fluid viscosity coefficient; k is the permeability coefficient; K is the permeability; ρ is the fluid density; η is the fluid dynamic viscosity coefficient; and g is the acceleration due to gravity.

From Formula (1.1), it can be seen that the influence of temperature on the flow characteristics mainly comes from its effect on the dynamic viscosity coefficient and fluid density. Due to the increase in temperature, the kinetic energy of the fluid increases, resulting in a decrease in viscosity coefficient. In addition, the surface tension of pore water also decreases with temperature, making it easier for the fluid to penetrate between soil particles. The change in fluid density will cause a change in porosity, resulting in a change in saturation and thus a change in permeability. Romero et al. [24] experimentally studied the relationship between permeability coefficient and temperature under fixed porosity and moisture content conditions.

$$\left.\frac{k_{\mathrm{w}}(e, w, T)}{k_{\mathrm{w}}(e, w, T_{\mathrm{r}})}\right| = \frac{\rho_{\mathrm{w}}(T)\eta_{\mathrm{w}}(T_{\mathrm{r}})}{\rho_{\mathrm{w}}(T_{\mathrm{r}})\eta_{\mathrm{w}}(T)} \approx 1 + \beta_{\mathrm{T}}(T - T_{\mathrm{r}}). \tag{1.2}$$

Among them, k_{w} is the permeability coefficient; e is the porosity ratio; w is the moisture content; T_{r} is the reference temperature; and β_{T} is the fitting coefficient.

Van Genuchten [43] proposed an expression for the relative permeability coefficient of unsaturated soil that is related to saturation

$$k_{\mathrm{r}} = \frac{K}{K_{\mathrm{s}}} = s_{\mathrm{e}}^{\frac{1}{2}}\left[\frac{\int_0^{s_{\mathrm{e}}}\frac{\mathrm{d}x}{\psi(x)}}{\int_0^1\frac{\mathrm{d}x}{\psi(x)}}\right]^2. \tag{1.3}$$

Among them, k_{r} is the relative permeability coefficient; k_{s} represents the permeability coefficient of saturated soil; K is the permeability coefficient of unsaturated soil; ψ is the matric suction of soil; and s_{e} is the effective moisture content. This formula can also be expressed as:

$$s_{\mathrm{e}} = \frac{\theta - \theta_{\mathrm{r}}}{\theta_{\mathrm{s}} - \theta_{\mathrm{r}}}. \tag{1.4}$$

Among them, θ is the volumetric water content; θ_{s} is the saturated moisture content; and θ_{r} is the residual moisture content.

Van Genuchten [43] also provided the relationship between effective saturation (s_{e}) and matric suction (ψ), known as the famous vG model.

$$s_{\mathrm{e}} = \left[\frac{1}{(\alpha\psi)^n + 1}\right]^{\frac{n-1}{n}}. \tag{1.5}$$

In the equation, α and n are fitting parameters that are related to the soil water characteristic curve (SWCC) of unsaturated soil.

Grant and Salehzadeh [74] proposed a model to describe the relationship between suction and temperature.

$$\frac{s(w, T)}{s(w, T_{\mathrm{r}})} = \left(\frac{a_1 + b_1 T}{a_1 + b_1 T_{\mathrm{r}}}\right)^{b_1}. \tag{1.6}$$

Among them, s is the matric suction force; w is the moisture content; T_{r} is the reference temperature; and a_1 and b_1 are the empirical coefficients.

Romero et al. [24] conducted permeability tests on unsaturated clay at different temperatures and found that the permeability coefficient of unsaturated soil increases with temperature, and the greater the saturation, the more significant this effect becomes. When fully saturated, the ratio of the permeability coefficients of 80–22 °C is about 1.3, and when the saturation is less than 75%, the ratio of the two approaches 1. Zhang et al. [36] also reached a similar conclusion by measuring the permeability of bentonite and sand mixtures at different temperatures. The results showed that the permeability coefficient of the soil sample slightly increased with increasing temperature, but the effect of temperature on the permeability coefficient was not significant when the saturation was low.

1.2.2 Experimental Study on Thermal Consolidation of Soil

Campanella and Mitchell [22] conducted cyclic temperature tests on clay, which showed that there was a small volume expansion during the cooling process, but the volume shrinkage and expansion were particularly small compared to the heating process. After one heating and cooling cycle, the overall volume shrinkage was observed, indicating irreversibility of the thermal volume change. As shown in Fig. 1.8, the clay sample was subjected to a thermal cycle of 20–90–20 °C under a confining pressure of 160 kPa. Heating to 90 °C caused volume shrinkage, but it did not recover during subsequent cooling. Therefore, the volume shrinkage caused by heating is irreversible.

Demars and Charles [44] found that the degree of volume change increases with the increase of clay plasticity and is independent of consolidation pressure, as shown in Fig. 1.9. When the soil sample undergoes temperature cycling and then increases the consolidation pressure, it exhibits overconsolidation characteristics [45, 46]. Many scholars have conducted experiments on the influence of temperature on the mechanical properties of clay. Towhata et al. [47] found that heating dry clay to

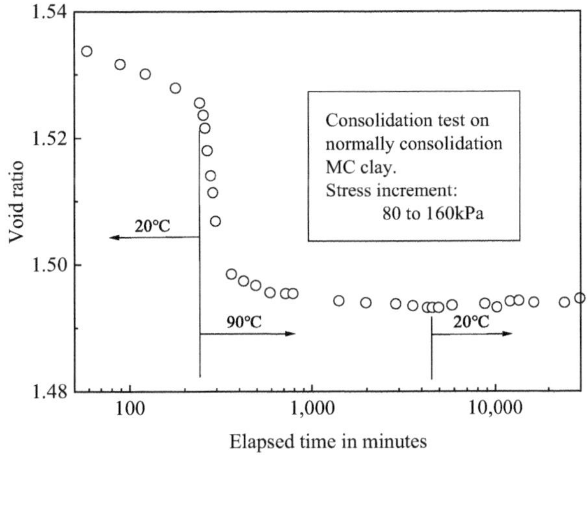

Fig. 1.8 Influence of cyclic temperature on MC clay volume change

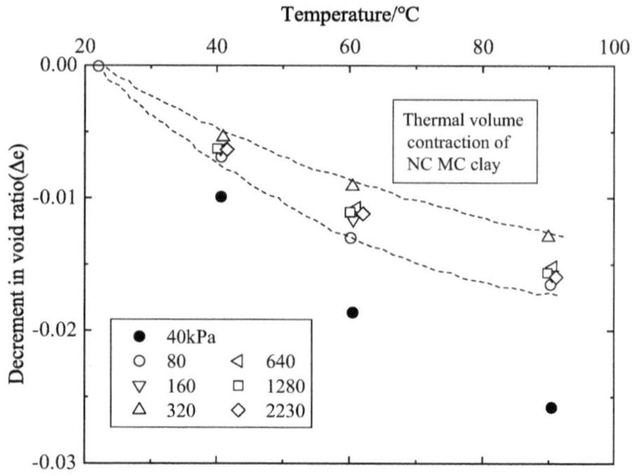

Fig. 1.9 Decrement of void ratio of MC clay due to heating

200 °C does not affect its properties, and heating wet clay should be used to study the effect of temperature on its properties.

As mentioned earlier, some scholars [11, 15, 45, 48] found that when the OCR value is low, heating causes thermal shrinkage, while at high OCR values, heating causes thermal expansion, as shown in Fig. 1.10.

Meanwhile, the study also found that the thermal shrinkage of normally consolidated soil samples is not related to the average effective stress value; when OCR value decreases, thermal shrinkage increases. When OCR = 1, only shrinkage occurs. The slope of body strain during the cooling stage is independent of the average effective stress. The excessive temperature between thermal contraction and thermal expansion decreases with the decrease of OCR value.

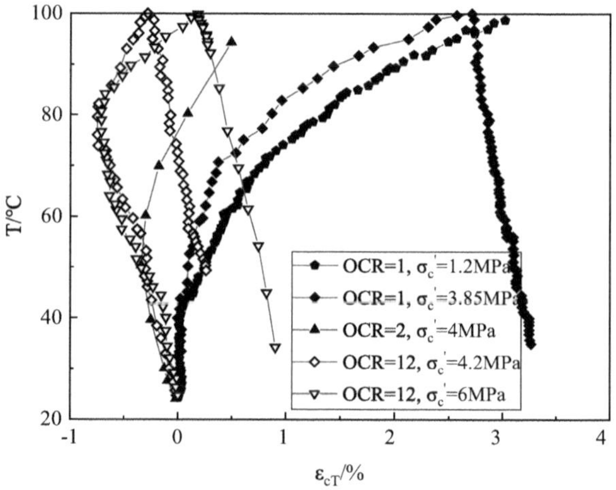

Fig. 1.10 Thermal volumetric changes of Boom clay samples at different OCR values

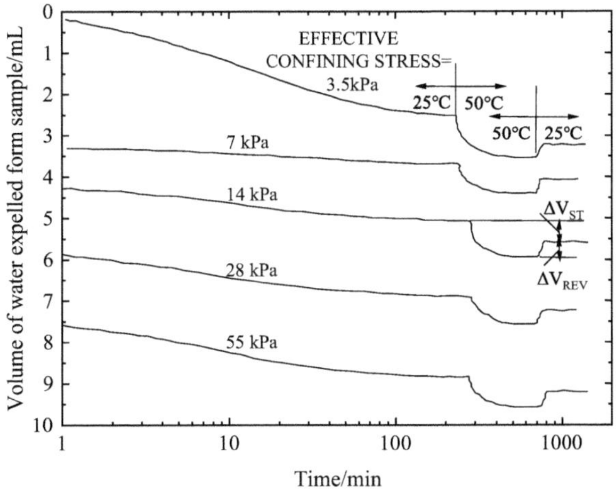

Fig. 1.11 Volume change–time curves for consolidation at several effective confining stresses in test

Demars and Charlie [49] studied the effect of cyclic temperature on soil volume change in isotropic consolidation tests. The study selected six different natural soil samples ranging from sand soil to high plasticity clay. Under a certain confining pressure, the soil sample is subjected to a heating–cooling cycle, with normal consolidation and overconsolidation stresses. The experimental results indicate that the reduction of porosity in normally consolidated soil samples due to temperature cycling is not related to effective confining stress, but to the plasticity of the soil sample, as shown in Fig. 1.11. For overconsolidated soil samples, the volume change is a function of the effective stress reduction or overconsolidation ratio, which is influenced by the cyclic temperature.

Eriksson [32] obtained the variation of preconsolidation pressure through isothermal consolidation tests on cohesive soil samples (plasticity index) at different temperatures, and the points obtained from different tests at different temperatures have variability. The average preconsolidation pressure obtained at 5 °C is 60 kPa, and it shows a nonlinear change with increasing temperature, as shown in Fig. 1.7. Hueckel and Baldi [29] defined the yield trajectory of temperature to pressure using this variation, indicating that the elastic space decreases with increasing temperature.

Hueckel and Baldi [29] conducted experimental research on the thermal plasticity of saturated clay and analyzed the experimental results of three types of clay samples based on the proposed soil thermal plasticity constitutive equation. The drainage test results under constant confining pressure show that the elastic zone has a strong dependence on temperature, and the thermal sensitivity of the elastic zone is different between normally consolidated soil and super consolidated soil. Under constant confining pressure, the normally consolidated soil sample undergoes thermoplastic strain hardening during drainage heating. The results of triaxial compression tests at constant temperature indicate an increase in soil ductility. When the principal stress remains constant, heating increases the pore water pressure while reducing the effective stress, and the soil sample ultimately fails at the critical line.

The temperature has little effect on the strength of soil samples. The strength of overconsolidated clay decreases with increasing temperature, while the strength of normally consolidated clay may increase. A plastic flow rule was proposed by comparing the experimental results of heating under constant confining pressure and loading under constant temperature.

Tsuchida et al. [46] found that the mechanical properties of clay were similar to those of lightly aged soil samples by consolidating it at high temperatures and cooling it down after consolidation. Therefore, they concluded that heating and aging have the same effect on clay samples. They mentioned that aging is caused by bonding, and heating accelerates this bonding process. Other researchers have also reached the same conclusion [15, 50].

Towhata et al. [15] found that volume of remolded kaolin ($I_p = 19$) shrinks when heated to 90 °C under a confining pressure of 160 kPa and then exhibits a slight overconsolidation effect under isothermal loading conditions at 90 °C. Plum and Esrig [45] observed that plastic shrinkage occurred in illite soil heated to 50 °C under a confining pressure of 110 kPa in a normally consolidated state. Subsequently, the soil was cooled to 24 and then loaded to 210 kPa for the same temperature cycle (24–50–24 °C), resulting in overconsolidation effect. After each cycle, the compression curve remains parallel to the previous curve, but due to the overconsolidation effect, it shifts to the right, proving the hardening effect. Demars and Charles [49] found that six different natural marine soils undergo normal consolidation under constant effective loads, and the irreversible reduction in thermal volume generated by thermal cycling (25–50–25 °C) is independent of the magnitude of the load. For high plasticity soils, the volume reduction is greater, and the high consolidation ratio results in smaller soil strain, which is consistent with the conclusion drawn by Paaswell [48].

Towhata et al. [15] conducted experimental research on the effect of temperature rise on the volume change of clay. They used a traditional consolidation apparatus with a heater and constant temperature chamber to study the interaction between clay particles and pore water. The consolidation test of normal soil samples shows that heating causes volume shrinkage, and when stress is continued, the soil sample exhibits overconsolidated clay. When the soil sample is heated for secondary compression, volume shrinkage will also occur. Although reloading the overconsolidated soil sample showed the same volume change, the soil sample expanded when the stress decreased. Finally, a hypothesis is proposed to explain the mechanism of thermal volume change of clay in the study, which is as follows:

(1) Each adsorbed water molecule has kinetic energy, and when the kinetic energy is large enough, there is a chance to escape the range of electric attraction and become free water.
(2) When the adsorbed water runs out, the thickness of the water layer decreases, resulting in clay particles being able to approach or combine with each other. Therefore, macroscopically, even when there is no pore water pressure, the volume of clay shrinks (secondary compression), indicating that this particle movement is irreversible.

(3) Heating increases molecular kinetic energy and accelerates the rate of volume contraction.
(4) The structure of clay particles generated by heating is denser or has more particle binding than before, making them harder than before heating.
(5) The shrinkage of the water absorbing layer leads to an increase in the cross-sectional area occupied by free water (even if the porosity remains constant), along with a decrease in viscosity. The expansion of the water flow channel results in accelerated seepage and volume changes (expansion or contraction).

Kuntiwattanakul et al. [34] studied the undrained shear characteristics of normally consolidated clay and over consolidated clay under high temperature (up to 90 °C). Research on isotropic consolidated undrained triaxial testing equipment based on controllable temperature, using a sequence of heating and consolidation to obtain the effect of temperature on clay. Research has shown that: (1) In the normal consolidation state, the initial secant modulus and shear strength increase with increasing temperature; (2) in the over consolidated state, only the initial secant modulus increases with temperature, and the shear strength is not significantly affected by temperature. And further explained the experimental results based on micro resistance and micro resistance of clay structure.

Akagi and Komiya [50] conducted a series of constant strain rate consolidation tests on remolded saturated clay. The results indicate that the consolidation size at high temperatures is independent of the strain rate, and the compression index is higher than that at room temperature. The consolidation coefficient (c_v) obtained at high temperatures is higher than that at room temperature, due to the decrease in viscosity of the soil skeleton and pore water. The volume compression coefficient (m_v) remains almost unchanged in the loading test, and the temperature induced changes are minimal.

Burghingoli et al. [21] conducted experimental research on the thermal properties of clay. Experiments were conducted on remolded clay and natural clay using a temperature-controlled triaxial apparatus, with temperature variations ranging from 20 to 60 °C. We separately studied the thermal properties of clay and the influence of temperature on its mechanical properties. The analysis of non-isothermal drainage and non-drainage test results determined the relationship between the temperature and time-dependent characteristics of the soil skeleton during the thermal solidification process. Therefore, for the creep process, the response of the soil is not only affected by thermal loads, but also depends on the stress history of the soil. The change in pore ratio of soil particles after readjustment is not only affected by temperature amplitude, but also depends on stress history, heating history, duration of heating, and the holding time of the highest temperature. Within the temperature range of the experiment, the deformation and strength of the soil are relatively independent of temperature. Thermal cycling causes soil hardening, and hardening effects only occur within a limited range of stress and deformation increments. Beyond this range, the characteristics of the soil skeleton are no longer affected by temperature history.

Delage et al. [14] studied the thermal solidification of clay. When saturated clay is heated, just like the soil around the disposal of nuclear waste deep in the formation, the thermal expansion of its composition produces pore water pressure, which changes its mechanical properties due to the influence of pore pressure and temperature gradient. By accurately measuring the volume change through slow heating tests, the heating changes of saturated clay were studied to predict soil reactions. The thermal solidification of clay was analyzed through rapid heating tests, and the results showed that hydraulic and heat transfers were separated. The rapid heating test results showed that due to the increase in permeability and the decrease in porosity, the consolidation coefficient did not change significantly with increasing temperature. The variation of permeability with temperature is determined by measuring the constant head at different temperatures, and the permeability coefficient value is obtained by considering the variation of free water viscosity with temperature. The relationship between permeability coefficient and porosity is independent of temperature, confirming that the flow in permeability tests is only related to free water. Delage et al. believed that thermal strain is essentially reversible after cooling and is related to the thermal shrinkage of saturated soil solid particles and adsorbed water. Due to the different thermal shrinkages of water and solid particles, the cooling rate should be sufficiently slow compared to heating to expel water. The thermal contraction coefficient of water ($\alpha_w = 10^{-5}$) is approximately 15 times greater than that of solid particles ($\alpha_s = 1.7 \times 10^{-4}$/°C), and the volumetric heat capacity of water ($C_w = 4.18 \times 10^6$) is nearly twice that of solid particles ($C_s = 2.856 \times 10^6$ J/(°C m^3)).

Sultan et al. [51] experimentally studied the influence of temperature on the preconsolidation pressure of clay and the effect of overconsolidation ratio on thermal volume change. The OCR value is defined by isotropic stress rather than the commonly used vertical stress, and the experiment is conducted in a temperature-controlled isotropic compression instrument. The experimental results indicate that the thermoplastic strain of normally consolidated soil is independent of the applied stress, and an exponential expression for the relationship between preconsolidation pressure and temperature is proposed. Based on the experimental results, the existing thermoplastic model was verified. In order to comprehensively explain the influence of OCR values on thermal strain, a plastic mechanism–thermal yield curve was proposed. Through experiments, it was found that the cooling rate has a significant impact on the slope of the temperature strain cooling curve, and a slower cooling rate should be adopted. Figure 1.12 shows the results of cooling the clay sample from 100 to 25 °C under a confining pressure of 4 MPa. There are three cooling rates, with 0.1 °C per 15 min for 100–78 °C, 0.1 °C per 10 min for 78–42 °C, and 0.1 °C per 8 min for 42–25 °C. The experimental results indicate that the cooling rate has a significant impact on the slope of the temperature strain curve. In this regard, some experimental results in literature suggest that the phenomenon of cooling expansion, especially in cases of rapid cooling, should be carefully considered.

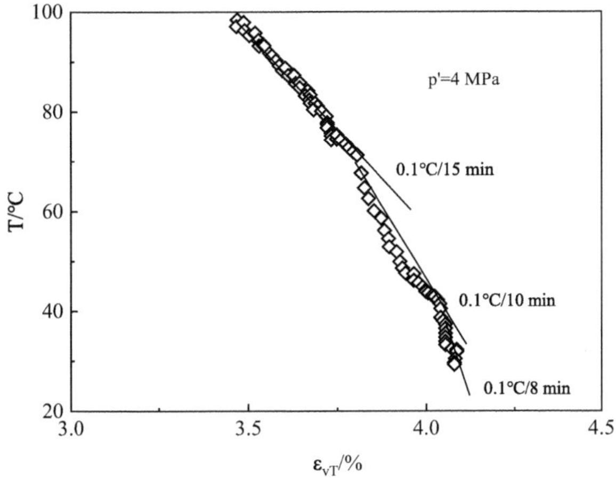

Fig. 1.12 Effect of the cooling rate on the thermal volume–change curve for Boom clay

Cekerevac and Laloui [13] studied the experimental results of the effect of temperature on the mechanical properties of saturated clay. A temperature-controlled three-axis device is used for the experiment, using CM clay material and applying a temperature of 22–90 °C. The experimental content includes: triaxial shear tests of soil samples with different overconsolidation ratios at room temperature and high temperature; consolidation tests at room temperature and high temperature; and heating and drainage tests on soil samples with different overconsolidation ratios. The experimental results provide the thermal properties of clay and compare the experimental results at 90 °C and room temperature. Based on the comparison results, the influence of heating on the thermal properties of clay is presented, mainly including: thermal volume change of heated drainage; the influence of temperature on the preconsolidation pressure and compression index; thermal effects on shear strength and critical state; the thermal effect of elastic modulus. In addition, thermal yielding was discussed and the evolution of yield limit was presented; The plastic strain changes caused by heating were discussed, and the differences in thermal properties of clay at room temperature and high temperature were presented.

Abuel-Naga et al. [41] studied the effect of heating up to 90 °C on the thermal properties of natural soft Bangkok clay. The direct effect of heating on volume change under different temperature and stress conditions was experimentally studied, and it was found that the volume change is related to the stress history but not to the stress level. The thermal evolution of preconsolidation pressure is nonlinear, and the percentage of evolution is independent of the stress level. The overconsolidation effect caused by the temperature rise cooling cycle, where the increase in temperature leads to a decrease in liquid viscosity and an increase in permeability. The experimental results of soft clay provide additional useful thermodynamic property and summarize the thermodynamic property of saturated clay compared to other types of clay. The thermal properties of saturated fine-grained soil at temperatures below the boiling point of pore water are attributed to physical chemical changes, soil plasticity, and the percentage of fine particles, which can also serve as indicators of thermal change intensity.

Abuel-Naga et al. [12] presented experimental results on the engineering properties of soft clay heated from room temperature 25 to 90 °C. The article provides a detailed introduction to an improved triaxial consolidation testing device that can be heated to 100 °C. Within the temperature range of the experiment, the thermal volume change exhibited by soft clay mainly depends on the stress history and is independent of the applied stress. The magnitude of the volume change is related to the plasticity index of the soil. As the temperature increases, the loading elastic zone decreases, resulting in hardening effect and increased permeability due to the decrease in liquid viscosity. Normally consolidated soil exhibits an over consolidated state after undergoing temperature rise and drop cycles, and its shear strength and secant modulus increase with increasing temperature. The experimental results can also enhance the understanding of the concept of thermal hydraulic characteristics of saturated soil.

Cui et al. [52] studied the temporal correlation of saturated clay in triaxial consolidation tests under different isotropic stresses and temperatures. Research has found that when the loading rate is less than 0.2 kPa/min at room temperature and the applied load is less than the preconsolidation pressure, the volume of the soil sample hardly changes, indicating the complete dissipation of pore water pressure. When the applied load and heating rate are high, the volumetric strain undergoes significant changes, especially when the applied load is much greater than the initial consolidation pressure, as shown in Fig. 1.13. Due to the low permeability of clay, complete consolidation takes a long time and it is difficult to distinguish the changes in volumetric strain over time between consolidation and creep.

Ghabezloo and Sulem [53] proposed the theoretical basis for the thermal response of liquid saturated porous materials under undrained conditions. It has been proven that the phenomenon of thermal pressurization is controlled by the difference in

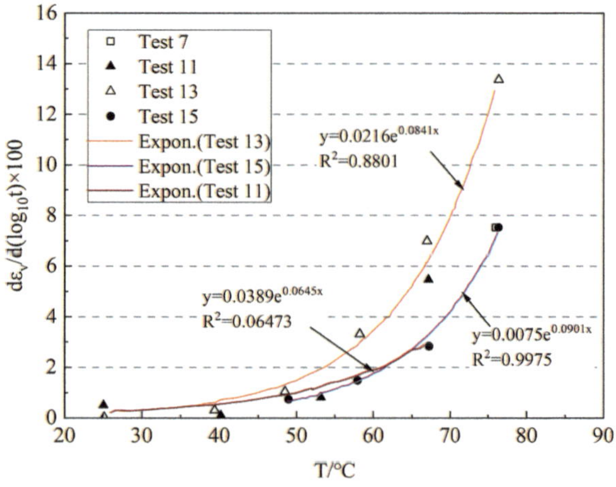

Fig. 1.13 Consolidation rate versus temperature for different heating tests on Boom clay

thermal expansion between pore fluids and solids, the stress dependence of compressibility, and the inelastic volume change of porous media. It is necessary to consider the influence of volume changes caused by other factors in the drainage system to demonstrate the undrained thermal pore elastic characteristics of saturated porous media in traditional triaxial apparatus. A simple modification scheme is proposed to accurately measure changes in pore pressure and volumetric strain under undrained heating. The experiment shows that the porosity of the material, the compressibility of the drainage system, and the volume change rate of the sample drainage system all affect the error of the drainage system.

Gulgun [54] studied the effects of heat treatment on the particle size, moisture content, specific gravity, plasticity, activity index, expansion and compression index, and mechanical properties of kaolin and bentonite. Firstly, the characteristics of kaolin and bentonite were tested using XRD and TG standard methods. Subsequently, all screened soil samples were heated to different temperatures (100, 200, 400, 500, and 600 °C) and maintained at the highest temperature for 24 h. Use triaxial compression test to determine the effect of heat treatment on the mechanical properties of all soil samples. Research has found that kaolin becomes non-plastic at 400 °C and bentonite becomes non-plastic at 500 °C, with particle size increasing with increasing heat treatment temperature. When the temperature increases to 400 °C, the expansion characteristics of kaolin and bentonite decrease significantly. There is a strong correlation between the shear resistance angle and the heat treatment of the two materials, and a correlation coefficient is given. The results show that kaolin and bentonite can be used as stable materials after heat treatment of 400 °C and 500 °C, respectively.

Shariatmadari and Saeidikam [55] studied the expansion and compression characteristics of bentonite sand mixtures at high temperatures using an improved consolidation apparatus. Use a comprehensive calibration program to calibrate the effect of high temperature on measured values. Research has shown that in a 1:1 bentonite sand mixture, an increase in temperature from 25 to 90 °C results in a decrease of approximately 20% in free expansion and strain. Due to the increase in permeability, the time required for expansion equilibrium decreases at high temperatures, which leads to an increase in compression rate and compression amount. According to some path studies, thermal and stress history are important parameters for accurately predicting the compressibility and swelling strain of expansive soil sand mixtures.

Vega et al. [56] described an experimental study on the effect of heating rate on the volume change of normally consolidated saturated loess caused by temperature using a temperature-controlled consolidation instrument with single-sided drainage. The results indicate that the heating rate has no significant effect on the thermal volume change after heating, but has a certain impact on the pore water pressure and maximum instantaneous shrinkage generated. Due to the dissipation of pore water pressure, soil samples with faster heating rates initially shrink larger than those with slower heating rates. After the pore pressure dissipates, the expansion rate of the soil sample is the same as the heating rate.

Tsutsumi and Tanaka [57] conducted constant strain rate load tests on remolded clay samples in the temperature range of 10–50 °C to investigate the effects of strain rate and temperature on the consolidation characteristics of clay. Under normal strain rates, for example, 10^{-6} s^{-1}, the influence of temperature on consolidation characteristics has been confirmed. Due to the decrease in viscosity of pore water at high temperatures, it is concluded that high-temperature conditions will lead to an increase in permeability coefficient. Due to the viscous properties of the soil skeleton, it has also been found that the yield consolidation pressure decreases with increasing temperature, and earlier scholars have reached the same conclusion [32, 50, 58, 59]. However, at high temperatures and low strain rates, clay samples exhibit different viscous behavior from typical models, such as the constant velocity line model, where the slope of the stress–strain curve is significantly reduced. The reason why the isokinetic model exhibits different characteristics under high temperature and low strain rates may be attributed to the formation of new soil structures to prevent external deformation [15, 46].

Cui et al. [60] studied the chemical thermal hydraulic properties of geological and engineering barriers and listed the characteristics of natural clay and compacted bentonite materials. For natural clay, research has found that the compression indices measured by consolidation instruments and isotropic compression tests are similar, and the compressibility of natural clay from MOL region is higher than that from Essen region. The shear strength of natural clay in MOL area is also higher than that in Essen area, indicating a significant influence of carbonate content. The overconsolidation ratio has a significant impact on the thermal volume change, with low overconsolidation ratios causing thermal shrinkage and high overconsolidation ratios causing thermal expansion. Time also affects the change characteristics of volume, and the time-dependent characteristics depend on the stress level and temperature. When the content of montmorillonite is relatively high, the salt content of pore water has a significant impact on the volume change characteristics. For bentonite materials, research has found that thermal shrinkage also occurs at low overconsolidation ratios, but it is related to suction, which promotes thermal expansion. Under constant volume conditions, humidification causes the permeability coefficient to first decrease and then increase.

Ghembaza et al. [18] experimentally studied the isotropic path thermal mechanical properties of sandy clay. The liquid limit of the material (W_1) used is 38%, and the plasticity index (I_p) is 19%. For a soil sample with a preconsolidation pressure of 100 kPa, a thermal consolidation test was conducted under constant stress, followed by a consolidation compression test at the corresponding temperature. The results of the heating test measuring volume change provide an equivalent relationship between thermal solidification and force solidification. The force consolidation at different temperatures indicates that the thermal compression modulus decreases at high temperatures (80 °C). In addition, permeability increases with temperature.

Alrtimi and Rouainia [61] proposed a stable instrument for measuring the thermal conductivity of soil. The design of the instrument is based on the application of Fourier's law, which generates one-dimensional uniform heat flux from two identical soil samples, forming a measurable temperature gradient to calculate the thermal conductivity of the pattern. Use a heat shield to detect and minimize heat loss. Research has found that non-radial heat loss conditions can be obtained through the heat shield temperature, which is close to the average of the surrounding temperature and the average soil temperature. All parameters that affect the measured values have been tested, and the results show good performance with an error margin of 5%. The application of the new testing program has conducted several pilot tests on undisturbed and remolded soil samples, demonstrating the simplicity of this device in measuring the thermal conductivity of soil under different conditions.

1.2.3 Theoretical Study on Thermal Consolidation of Soil

Hueckel and Baldi [29] first proposed an elastic–plastic thermodynamic constitutive model for saturated soil by extending the modified Cambridge model. The yield surface at different temperatures can be represented by the following equation:

$$f = \left(\frac{2p'}{p'_c} - 1\right)^2 + \left(\frac{2q}{Mp'_c}\right)^2 - 1 = 0. \tag{1.7}$$

Among them, p'_c the maximum preconsolidation effective stress controls the size of the yield surface, and as the temperature increases, the yield surface will shrink. The LY (Load Yield) yield curve is used to describe the variation of the yield surface with temperature.

$$p'_c = p_{atm}\left(\frac{e_1 - (1 - \alpha_0\Delta T)\left[e_g - (1 + e_0)\varepsilon_v^{tp}\right]}{\lambda - K_T}\right). \tag{1.8}$$

Among them, λ and K_T are respectively the isothermal elastoplastic modulus and elastic modulus; e_0 and e_1 are the initial porosity ratio and the porosity ratio in the reference state, respectively; and α_0 is the corresponding temperature influencing parameters.

Due to the influence of temperature on the elastic range, the criteria for distinguishing loading and unloading differ from traditional elastic–plastic theory, which is not only related to stress paths but also to changes in temperature. Based on this theoretical framework, some scholars [16, 29, 31] have also proposed thermal coupling models for saturated soils, and these improved models further enhance the ability of the models to describe thermal characteristics. Cui et al. [16] proposed a TY (Thermal Yield) yield surface considering thermal yielding, based on thermoplasticity theory and experimental studies. It forms an elastic zone in the T–p' plane with the LY yield surface. The TY yield surface is related to the maximum temperature that the soil has been subjected to. Due to its introduction, the model can better simulate the thermal coupling characteristics of soil. Compared with the model proposed by Hueckel and Baldi [29], it has the following improvements: heating normally consolidated soil under constant pressure can cause overconsolidation effect; heating of overconsolidated soil will result in plastic deformation. But as the yield surface increases, the model parameters also increase accordingly, making the solution more complex.

Abuel-Naga et al. [31] proposed a thermomechanical coupling model under critical conditions, and the expression for the yield surface is as follows:

$$\frac{p'}{p'_c} = \frac{M^2 + \alpha^2}{M^2 + (\eta - \alpha)^2}. \tag{1.9}$$

α is a temperature-dependent parameter. This expression can reflect the influence of temperature on the yield surface, and improve the model's ability to simulate shear characteristics.

In addition, He et al. [62] studied the changes in temperature, pore pressure, and thermal stress around a spherical cavity in an infinitely large saturated porous medium using non-equilibrium thermal porous elasticity theory and obtained asymptotic solutions for temperature, pore pressure, and thermal stress through Laplace transform. Hong et al. [63] compared three well-known thermodynamic models [16, 17]. Zhou and Ng [64] proposed a thermomechanical coupling model for saturated soil based on the boundary surface plasticity theory, considering the influence of temperature on the size and shape of the boundary interface, which can simulate the characteristic of reduced shear modulus under small deformation.

There have been many studies on the THM coupling problem in soil from three aspects: theoretical analysis, experiments, and numerical simulations. Early THM coupling problems were usually analyzed using porous media theory, including Biot isothermal consolidation theory, classical elastoplastic theory, and mixture theory. Many scholars have proposed THM coupling models for saturated or unsaturated soils based on these theories and different research backgrounds. Gatmiri and Delage [65] proposed a THM-coupled model that takes into account the nonlinear elastic deformation, compressibility, thermal expansion, and thermal convection of soil media. Wu et al. [66] studied the effect of temperature on the hydraulic properties of unsaturated soils and found that the maximum yield curves of preconsolidation pressure and suction would decrease with increasing temperature, i.e., thermal softening. They proposed a THM-coupled model for unsaturated soils. Tong et al. [67] proposed a THM coupling model for unsaturated soils based on mixture theory, considering solid-phase deformation, liquid-phase and gas-phase transport, and heat conduction. Dumont et al. [68] proposed a simple THM coupling model for unsaturated soils using the concept of effective stress and fewer parameters. Cai et al. [69] proposed a fully coupled theoretical framework for unsaturated soil THM based on continuum mechanics and mixture theory.

The numerical calculation methods for THM coupling problems mainly include finite element method, finite difference method, discrete element method, and SPH calculation method. Wang et al. [70] used the parallel Newton–Raphson algorithm with domain decomposition technique to solve the THM coupling problem in porous media with two-phase flow of water and gas. This algorithm considers subregions as composite units, and the equations of each subregion only consider the degrees of freedom of its boundary nodes. As long as the unknown variables of the boundary nodes of the subregion are determined, the internal variables can be independently solved. Rutqvist et al. [71] proposed the governing equations for the THM coupling process in porous media and compared and analyzed the frameworks of four finite element calculation methods. They found that the main difference lies in the treatment of the gas phase. Li et al. [72] used a mixed finite element method to perform THM-coupled numerical analysis on porous media elastoplastic materials. Lu et al. [73] used the finite difference method to simulate the THM coupling problem of clay barriers and studied the effect of radioactive waste heat release on soil deformation. Analysis suggests that due to the complex geometric and physical state of clay, it is very difficult to simulate it using finite difference method. It is recommended to combine it with finite element method for simulation calculations.

1.3 Summary

This section provides a detailed introduction to the current research status of soil thermal consolidation, including theoretical and experimental studies. Through the summary of existing research results, it can be seen that valuable research achievements have been made in the field of multi-field coupling of soil, especially in the study of the influence of temperature field on soil mechanical properties, including experimental and theoretical model research, providing a solid foundation for further in-depth research. However, there are still some problems in current research, which are also new challenges facing the development of soil mechanics:

(1) At present, research on multi-field coupling of soil mainly relies on theoretical models, lacking corresponding experimental data support. The main reasons include three aspects: 1) lack of corresponding testing equipment; 2) considerable difficulties in sample preparation and test control; 3) relatively long test cycle. The specimen device is an essential support for conducting research. Most of the existing experimental devices used for multi-field coupling research are modified from the original consolidation or triaxial apparatus, which are relatively simple and require improvement in control accuracy. Therefore, the development of high-precision and easy to control experimental devices is also a current research requirement. Furthermore, as a semi-empirical and semi-theoretical discipline, soil mechanics relies on experiments as the foundation for further in-depth theoretical research. Therefore, it is essential to attach importance to experimental research in soil mechanics.

(2) In the study of thermal solidification of saturated soil, there is a lack of systematic experimental research, especially on the thermal deformation and pore pressure characteristics under different loads and temperature paths. Moreover, many experimental studies only conduct qualitative analysis, and there is relatively little research on the mechanism of thermal solidification. In addition, the complexity and diversity of influencing factors make the research not in-depth

enough. There are relatively few experimental studies on various complex stress and temperature paths, and the thermal solidification characteristics of soil under such complex paths are not clear enough.

(3) In the research of multi-field coupling models, the coupling characteristics are often reflected by adding coupling terms. However, in the corresponding constitutive equations, the coupling effects of other fields are rarely involved. The resulting coupling models are not truly multi-field coupling and do not fundamentally reflect the characteristics of multi-field coupling. In addition, some of the coupling coefficients involved in the model may not have specific physical meanings and are difficult to determine through experiments. Most existing soil theoretical models are based on traditional constitutive theories, such as the widely used classical elastoplastic mechanics theory. The yield criteria or hardening parameters involved are often fitted through some indoor test results, lacking a rigorous and scientific theoretical basis, and the resulting models may only have good simulation effects on a certain test result. Therefore, there is still a lack of a unified constitutive theory that is more universally applicable to geotechnical materials.

References

1. Krishnaiah S, Singh D (2003) Determination of influence of various parameters on thermal properties of soils. Int Commun Heat Mass Transfer 30(6):861–870
2. Abuel-Naga HM, Bergado DT, Bouazza A, Pender MJ (2009) Thermal conductivity of soft Bangkok clay from laboratory and field measurements. Eng Geol 105(3–4):211–219
3. Li H, Nagano K, Lai Y (2012) Heat transfer of a horizontal spiral heat exchanger under groundwater advection. Int J Heat Mass Transf 55(23–24):6819–6831
4. Brandl H (2006) Energy foundations and other thermo-active ground structures. Géotechnique 56(2):81–122
5. Lamarche L, Kajl S, Beauchamp B (2007) A review of methods to evaluate borehole thermal resistances in geothermal heat-pump systems. Geothermics 39(2):187–200
6. Man Y, Yang H, Diao N, Liu J, Fang Z (2012) A new model and analytical solutions for borehole and pile ground heat exchangers. Int J Heat Mass Transf 53(13–14):2593–2601
7. Mitchell JK, McMillan JC, Green SL, Sisson RC (1982) Field testing of cable backfill systems. In: Proceedings of the symposium on underground cable thermal backfill, Toronto, Canada, pp 19–33
8. Laloui L, Modaressi H (2002) Modelling of the thermo-hydro-plastic behavior of clays. In: International workshop on geomechanics-hydromechanical and thermohydromechanical behaviour of deep argillaceous rock, pp 161–170
9. Gens A, Olivella S (2006) Coupled thermo-hydro-mechanical analysis of engineered barriers for high level radioactive waste. Chin J Rock Mech Eng 25(4):670–680
10. Gens A (2010) Soil-environment interactions in geotechnical engineering. Géotechnique 60(1):3–74
11. Baldi G, Hueckel T, Pellegrini R (1988) Thermal volume changes of the mineral-water system in low-porosity clay soils. Can Geotech J 25(4):807–825
12. Abuel-Naga HM, Bergado DT, Ramana GV, Grino L, Rujivipat P, Thet Y (2006) Experimental evaluation of engineering behavior of soft Bangkok clay under elevated temperature. J Geotech Geoenviron Eng 132:902–910

13. Cekerevac C, Laloui L (2004) Experimental study of thermal effects on the mechanical behaviour of a clay. Int J Numer Anal Meth Geomech 28:209–228
14. Delage P, Sultan N, Cui YJ (2000) On the thermal consolidation of Boom clay. Can Geotech J 37:343–354
15. Towhata I, Kuntiwattanaul P, Seko I, Ohishi K (1993) Volume change of clays induced by heating as observed in consolidation tests. Soils Found 33:170–183
16. Cui YJ, Sultan N, Delage P (2000) A thermomechanical model for saturated clays. Can Geotech J 37(3):607–620
17. Abuel-Naga H, Bergado D, Bouazza A, Ramana G (2007) Volume change behaviour of saturated clays under drained heating conditions: experimental results and constitutive modeling. Can Geotech J 44(8):942–956
18. Ghembaza MS, Taibi S, Fleureau JM (2014) Thermo-hydro-mechanical behaviour of a sandy clay on isotropic paths. Eur J Environ Civ Eng 18(2):206–222
19. Tang AM, Cui YJ, Barnel N (2008) Thermo-mechanical behaviour of a compacted swelling clay. Géotechnique 58(1):45–54
20. Uchaipichat A, Khalili N (2009) Experimental investigation of thermo-hydromechanical behaviour of an unsaturated silt. Géotechnique 59(4):339–353
21. Burghignoli A, Desideri A, Miliziano S (2000) A laboratory study on the thermomechanical behaviour of clayey soils. Can Geotech J 37(4):764–780
22. Campanella RG, Mitchell JK (1968) Influence of temperature variations on soil behavior. J Soil Mech Found Div 94:709–734
23. Bai B, Shi XY (2017) Experimental study on the consolidation of saturated silty clay subjected to cyclic thermal loading. Geomech Eng 12(4):707–721
24. Romero E, Gens A, Lloret A (2003) Suction effects on a compacted clay under non-isothermal conditions. Géotechnique 53(1):65–81
25. Graham J, Tanaka N, Crilly T, Alfaro M (2001) Modified cam-clay modelling of temperature effects in clays. Can Geotech J 38(3):608–621
26. Uchaipichat A (2005) Experimental investigation and constitutive modelling of thermo-hydro-mechanical coupling in unsaturated soils. The University of New South Wales, Sydney
27. Ghahremannejad B (2006) Thermo-mechanical behaviour of two reconstituted clays. University of Sydney, Sydney
28. Zhou C (2014) Experimental study and constitutive modelling of cyclic behaviour at small strains of unsaturated silt at various temperatures. The Hong Kong University of Science and Technology, Hong Kong
29. Hueckel T, Baldi G (1990) Thermoplasticity of saturated clays: experimental constitutive study. J Geotech Eng 116(12):1778–1796
30. Pusch R, Güven N (1990) Electron microscopic examination of hydrothermally treated bentonite clay. Eng Geol 28(3):303–314
31. Abuel-Naga H, Bergado D, Bouazza A, Pender M (2009) Thermomechanical model for saturated clays. Géotechnique 59(3):273–278
32. Eriksson LG (1989) Temperature effects on consolidation properties of sulphide clays. In: Proceedings of the 12th international conference on soil mechanics and foundation engineering, Rio de Janeiro, pp 2087–2090
33. Cekerevac C, Laloui L (2010) Experimental analysis of the cyclic behaviour of kaolin at high temperature. Géotechnique 60(8):651–655
34. Kuntiwattanakul P, Towhata I, Ohishi K, Seko I (1995) Temperature effects on undrained shear characteristics of clay. Soils Found 35(1):147–162
35. Abuel-Naga HM, Bergado DT, Lim BF (2007) Effect of temperature on shear strength and yielding behavior of soft Bangkok clay. Soils Found 47(3):423–436
36. Burghignoli A, Desideri A, Miliziano S (1995) Discussion on volume change of clays induced by heating as observed in consolidation test. Soil Found 35(3):122–124
37. Towhata I, Kuntiwattanakul P, Seko I, Ohishi K (1995) Discussion on volume change of days induced by heating as observed in consolidation tests. Soils Found 35(3):124–127

38. Morin R, Silva AJ (1984) The effects of high pressure and high temperature on some physical properties of ocean sediments. J Geophys Res 89(Bl):511–526
39. Finn FN (1951) The effect of temperature on the consolidation of soils. Am Soc Test Mater Spec Tech Publ 126:65–72
40. Habibagahi K (1977) Temperature effect and the concept of effective void ratio. Indian Geotech J 7:14–34
41. Abuel-Naga HM, Bergado DT, Soralump S, Rujivipat P (2005) Thermal consolidation of soft Bangkok clay. Lowland Technol Int 7(1):13–21
42. Ye WM, Wan M, Chen B, Chen YG, Cui YJ, Wang J (2013) Temperature effects on the swelling pressure and saturated hydraulic conductivity of the compacted GMZ01 bentonite. Environ Earth Sci 68(1):281–288
43. van Genuchten MT (1980) A closed-form equation for predicting the hydraulic conductivity of unsaturated soils. J Soil Sci Soc Am 44:892–898
44. Demars KR, Charles RD (1981) Soil volume change induced by temperature cycling. Can Geotech J 19(2):189–194
45. Plum RL, Esrig MI (1969) Some temperature effects on soil compressibility and pore water pressure. Highway research board special report, pp 231–242
46. Tsuchida T (1991) Effect of aging of marine clay and its duplication by high temperature consolidation. Soil Found 31(4):133–147
47. Towhata I, Kuntiwattanaul P, Kobayashi H (1993) A preliminary study on heating of clays to examine possible effects of temperature on soil-mechanic properties. Soil Found 33(4):184–190
48. Paaswell RE (1967) Temperature effects on clay soil consolidation. J Soil Mech Found Div 93:9–22
49. Demars KR, Charles RD (1982) Soil volume changes induced by temperature cycling. Can Geotech J 19(2):188–194
50. Akagi H, Komiya K (1995) Constant rate of strain consolidation properties of clayey soil at high temperature. In: Compression and consolidation of clayey soils, pp 3–8
51. Sultan N, Delage P, Cui Y (2002) Temperature effects on the volume change behavior of Boom clay. Eng Geol 64:135–145
52. Cui YJ, Le TT, Tang AM, Delage P, Li XL (2009) Investigating the time dependent behaviour of Boom clay under thermo-mechanical loading. Geotechnique 59(4):319–329
53. Sulem J, Ghabezloo S (2010) Temperature induced pore fluid pressurization in geomaterials. Riv Ital Geotec 44(1):35–49
54. Gulgun Y (2011) The effects of temperature on the characteristics of kaolinite and bentonite. Sci Res Essays 6(9):1928–1939
55. Shariatmadari N, Saeidijam S (2011) The effect of elevated temperature on compressibility and swelling of bentonite-sand mixtures. Electron J Geotech Eng 137–146
56. Vega A, Coccia CJR, Tawati AE, Mccartney JS (2012) Impact of the rate of heating on the thermal consolidation of saturated silt. In: GeoCongress, pp 4476–4485
57. Tsutsumi A, Tanaka H (2012) Combined effects of strain rate and temperature on consolidation behavior of clayey soils. Soils Found 52(2):201–215
58. Boudali M, Leroueil S, Srinivasa Murthy BR (1994) Viscous behaviour of natural clays. In: Proceedings of the 13th international conference on soil mechanics and foundation engineering, New Delhi, India, pp 411–416
59. Marques M, Leroueil S, de Almeida M (2004) Viscous behaviour of St-Roch-de-l'Achigan clay Quebec. Can Geotech J 41:25–38
60. Cui YJ, Tang AM (2013) On the chemo-thermo-hydro-mechanical behaviour of geological and engineered barriers. J Rock Mech Geotech Eng 5:169–178
61. Alrtimi A, Rouainia M, Manning DAC (2014) An improved steady-state apparatus for measuring thermal conductivity of soils. Int J Heat Mass Transf 72:630–636
62. He LW, Jin ZH (2011) Effects of local thermal non-equilibrium on the pore pressure and thermal stresses around a spherical cavity in a porous medium. Int J Eng Sci 49:240–252
63. Hong P, Pereira J, Tang A, Cui YJ (2013) On some advanced thermo-mechanical models for saturated clays. Int J Numer Anal Meth Geomech 37(17):2952–2971

64. Zhou C, Ng CWW (2015) A thermomechanical model for saturated soil at small and large strains. Can Geotech J 52(8):1101–1110
65. Gatmiri B, Delage P (1997) A new formulation of fully coupled thermal-hydro-mechanical behavior of saturated porous media: numerical approach. Int J Numer Anal Meth Geomech 21(3):199–225
66. Wu W, Li X, Charlier R, Collin F (2004) A thermo-hydro-mechanical constitutive model and its numerical modelling for unsaturated soils. Comput Geotech 31:155–167
67. Tong FG, Jing LR, Zimmerman RW (2010) A fully coupled thermo-hydro-mechanical model for simulating multiphase flow, deformation and heat transfer in buffer material and rock masses. Int J Rock Mech Min Sci 47:205–217
68. Dumont M, Taibi S, Fleureau JM, Bekr NA, Saouab A (2010) Modelling the effect of temperature on unsaturated soil behavior. CR Geosci 342(12):892–900
69. Cai GQ, Zhao CG, Sheng DC, Zhou AN (2014) Formulation of thermo-hydro-mechanical coupling behavior of unsaturated soils based on hybrid mixture theory. Acta Mech Sin 30(4):559–568
70. Wang X, Gawin D, Schrefler BA (1996) A parallel algorithm for thermo-hydro-mechanical analysis of deforming porous media. Comput Mech 19(1):94–104
71. Rutqvist J, Börgesson L, Chijimatsu M, Kobayashi A, Jing L, Nguyen TS, Noorishad J, Tsang CF (2001) Thermohydromechanics of partially saturated geological media: governing equations and formulation of four finite element models. Int J Rock Mech Min Sci 38(1):105–127
72. Li XK, Liu Z, Lewis RW (2005) Mixed finite element method for coupled thermo-hydro-mechanical process in poro-elasto-plastic media at large strains. Int J Numer Meth Eng 64(5):667–708
73. Lu Y, Wu X, Cui Y (2010) On the thermo-mechanical properties of unsaturated soils. J Rock Mech Geotech Eng 2(2):143–214
74. Grant S, Salehzadeh A (1996) Calculation of temperature effects on wetting coefficients of porous solids and their capillary pressure functions. Water Resour Res 32(2):261–270

Chapter 2
Experimental Study on Thermal Consolidation Behavior of Saturated Silty Clay Under Non-isothermal Conditions

2.1 Overview of the Test

A self-developed temperature-controlled consolidation test apparatus was used to conduct indoor tests on the thermal consolidation characteristics of a saturated powdery clay. The test apparatus includes a temperature control system, a peripheral pressure system, a counterpressure control system, and a pore water pressure measurement system, and the thermal consolidation tests were carried out at temperatures ranging from 25 °C to 50 °C, in which the sample peripheral pressures were 50 kPa, 100 kPa, and 150 kPa, respectively. The evolutions of the temperatures, the pore water pressures, and the strain of the consolidation body with the peripheral pressures in the course of the thermal consolidation process were analyzed by means of the test phenomena, the evolution of pore water pressure and solidus strain with peripheral pressure during the thermal consolidation process was analyzed through the experimental phenomena, and the evolution of pore water pressure and solidus strain with peripheral pressure was explained by applying the study of thermal property mechanism.

2.1.1 Test Equipment

(1) Description of the Test Setup

Figure 2.1 shows the schematic diagram of the self-developed thermal consolidation test device. The top cover of the temperature-controlled solidification pressure chamber is connected to the side wall of the pressure chamber by welding; the side wall of the pressure chamber is connected to the base of the pressure chamber by fixing bolts, and the gap is sealed by an O-type rubber ring; the axial pressure rod passes through the center of the top cover through the water-stopping bushing to contact with the specimen cap; the specimen is placed on the permeable stone at

G. Yang and B. Bai, *Experimental and Theoretical Study on Thermal Consolidation Behavior of Saturated Clay*, https://doi.org/10.1007/978-981-96-6909-7_2

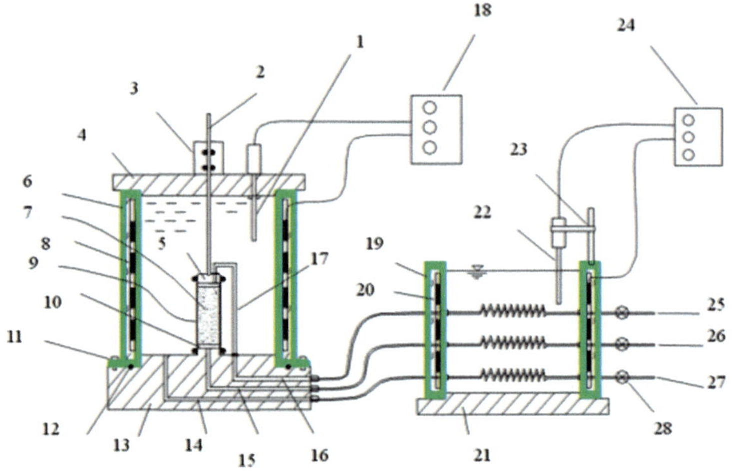

Fig. 2.1 Schematic diagram of test apparatus for thermal consolidation. *Note 1* Pressure chamber electric thermocouple, *2* axial pressure rod, *3* bushing, *4* top cover, *5* specimen cap, *6* pressure chamber side wall, *7* test, *8* pressure chamber cartridge thermocouple, *9* rubber membrane, *10* permeable stone, *11* fixing bolts, *12* O-ring rubber ring, *13* pressure chamber base, *14* peripheral pressure connection holes, *15* pore water pressure measurement holes, *16* test drain holes, *17* drainage pipe, *18* temperature regulator, *19* water vessel sidewall, *20* water vessel electric heating plate, *21* water vessel base, *22* water vessel electric heating couple, *23* bracket, *24* water vessel temperature controller, *25* test drain pipe, *26* pore water pressure measurement pipe, *27* perimeter pressure connection pipe, *28* valve

the center of the base of the pressure chamber and is connected to the top of the specimen with the permeable stone as well as to the specimen cap; the specimen is wrapped with a water-impermeable latex membrane; the two ends of the drain pipe are connected to the top of the specimen cap and the inlet of the specimen drain hole on the base of the pressure chamber, respectively.

The pressure chamber is a cavity structure made of stainless steel, the pressure chamber cylindrical heating plate is embedded in the side wall of the pressure chamber, the ends of the resistance wire are connected to the temperature regulator, and the outer side of the pressure chamber cylindrical heating plate is filled with asbestos for heat insulation. The pressure chamber electric thermocouple is mounted on the top cover, one end of which is placed in the water-filled pressure chamber, and the output end of which is connected to the temperature regulator.

The outlet of the specimen drainage hole is connected with the specimen drainage pipe through a connector; the outlet of the pore water pressure measurement hole is connected with the pore water pressure measurement pipe through a connector; the peripheral pressure connection hole is connected with the peripheral pressure connection pipe through a connector. The specimen drain pipe, pore water pressure measuring tube, and confining pressure connecting pipe are all made of stainless steel. The middle section features a helical structure and is placed in a water-filled container to facilitate efficient heat exchange with water. The specimen drain tube led from the constant temperature water container is connected with the body change tube; the pore water pressure measurement tube led from the constant temperature water container is connected with the pore water pressure measurement system; the peripheral pressure connection tube led from the constant temperature water container is connected with the stabilized pressure regulating system.

The body-variable tube, the pore water pressure measurement system, and the pressure stabilizing and regulating system all adopt the common supporting equipment of the conventional triaxial consolidation apparatus as shown in Fig. 2.2.

(2) Characteristics of Thermosetting Pressure Chambers This test device overcomes the defects of the existing conventional triaxial consolidation pressure chamber that cannot perform high temperature and changing temperature load tests and provides a temperature-controlled consolidation pressure chamber system with higher sensitivity to meet the needs of thermal consolidation tests of saturated soil specimens under different temperature loads

This test device overcomes the defects of the existing conventional triaxial consolidation pressure chamber that cannot perform high temperature and changing temperature load tests and provides a temperature-controlled consolidation pressure chamber system with higher sensitivity to meet the needs of thermal consolidation tests of saturated soil specimens under different temperature loads as shown in Fig. 2.3. The temperature inside the pressure chamber is controlled to vary within the range of 17–100 °C, and the test is determined according to the actual magnitude of the temperature load to be simulated on the soil specimen as well as the temperature change process. The pressure system provides a maximum pressure of 3 MPa, which is formed by a vacuum pump combined with an enclosure pressure connection pipe through a pressure stabilizing and regulating system. When the enclosure pressure in the pressure chamber is lower than the set pressure value, the vacuum pump automatically adjusts the enclosure pressure to the set value by means of the pressure regulating system. The counterpressure control system can provide and measure counterpressure up to 3 MPa. The pore water pressure measurement system is led by the pore water pressure measurement pipe through the pore water pressure measurement hole located at the base of the pressure chamber specimen and then connected to the pore water pressure sensor, and the pore water pressure value can finally be read on the display panel.

In the conventional triaxial consolidation test of saturated soil samples (room temperature and isothermal conditions), the volume change of the sample is equal to the amount of pore water discharged from the specimen, so the volume strain of the specimen can be known as long as the amount of pore water discharged is measured. However, under non-isothermal conditions, due to the thermal expansion or contraction of the pore water and the soil particles themselves, the volume change is difficult to be accurately determined by the above method. Indirectly determining the volume change of the specimen by measuring the volume change of water in the pressure chamber is a method that deserves attention. However, in the ordinary triaxial consolidation test, due to the thermal deformation of the test device itself (such as the Plexiglas pressure chamber, plastic drainage pipe, and rubber membrane), and due to the thermal expansion or contraction of the pore water and the soil particles themselves, the volume change of the specimen cannot be accurately determined directly by the amount of the pore water discharged, and the test results therefore have a large error. Satisfactory accuracy can be obtained by replacing components such as the pressure chamber and measurement piping with metal materials (Fig. 2.1). Specifically, the pressure chamber pressure control system can be used to measure the volume change of water in the pressure chamber, and the counterpressure system can be used to measure the volume change of pore water in the specimen. Through comparative analysis, the consistency of the two measurements under isothermal conditions can be verified, which in turn reflects the effect of thermal volume change of pore water and soil particles under non-isothermal conditions.

In addition, the accuracy of the measurement is also affected by the fact that the pressurization system and the drain system are at the laboratory temperature environment (17 °C), while the specimen and the water in the pressure chamber are at the test set temperature (e.g., 50 °C), which leads to a certain gradient change of the water temperature in the different parts of the drain pipe. Therefore, the metal tube connecting the pressure chamber to the pressure control and measurement system is placed in a thermostatic (laboratory temperature) container filled with water.

In this way, when the temperature in the pressure chamber changes, it does not affect the measurement of water volume change. Not only is it possible to accurately measure the thermal volumetric strain of the specimen and the pore water pressure, but it is also possible to measure the amount of pore water discharged from the specimen at the temperature of the test chamber.

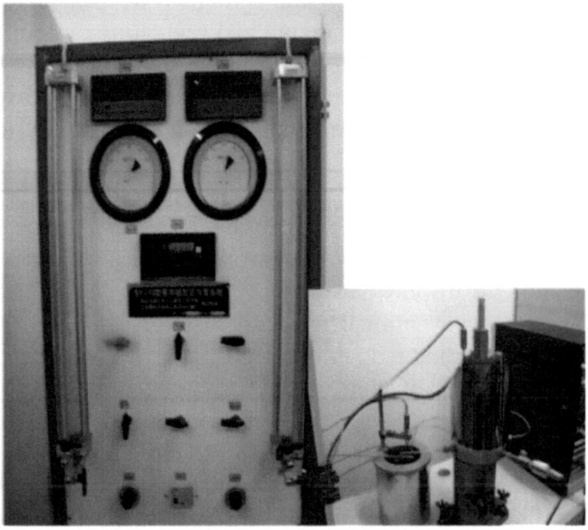

Fig. 2.2 Outside view drawing of test apparatus for thermal consolidation

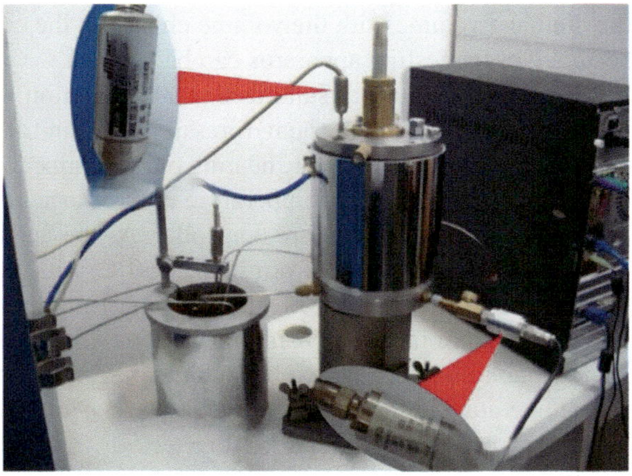

Fig. 2.3 Outside view drawing of consolidation pressure chamber

2.1.2 *Test Soil*

The test soil samples were selected from a kind of powdery clay, which was air-dried, crushed, and passed through 0.5 mm sieve to prepare remodeled soil samples. The basic physical property indexes of the soil were as follows: plastic limit ω_{p} was 17.6%, liquid limit ω_{L} was 31.4%, plasticity index I_{P} was 13.8, and relative density of soil particles G_{s} was 2.71.

Soil samples with a specified water content are prepared. The pre-weighed soil sample is divided into five layers and placed into a compaction mold, with each layer compacted 15 times. The sample is uniformly compacted into a specimen with a height of 8 cm and a diameter of 3.91 cm, controlling the dry unit weight γ_d at 15.5 kN/m^3.

2.1.3 *Test Program*

The test was set up with three sets of peripheral pressures of 50 kPa, 100 kPa, and 150 kPa, respectively. After installing the specimens, the pressure chamber was heated to 25 °C for isothermal consolidation. The temperature was slightly greater than the ambient temperature of the test chamber, which was 23 °C. Initial consolidation was considered to be completed when the change in pore water pressure was less than 1 kPa within 10 h. This process was considered to be the first step in the consolidation of the saturated soil. This process is the conventional triaxial consolidation test on saturated soil samples (room temperature, isothermal conditions); This process is a conventional triaxial consolidation test on saturated soil samples (under room temperature conditions). Theoretically, the volume change of the specimen should be equal to the amount of pore water discharged during this process. By comparing the drainage volume with the volume change in the confining pressure chamber, the accuracy of the testing apparatus can be verified.

After the specimen has stabilized in consolidation, carry out the undrained heating test. Close the valve of the counterpressure drain and adjust the temperature controller to 50 °C. A cylindrical electric heating plate is embedded in the cavity of the sidewall of the pressure chamber. Temperature control is achieved by connecting both ends of its resistor to a temperature controller. As the heating plate starts to operate, the temperature gradually rises from 25°C to 50°C, with the entire heating process lasting approximately 100 minutes. During this process, the corresponding time for every 1°C temperature increase (using the time at 25°C as the initial time), the readings from the confining pressure tube, and the pore water pressure readings are recorded. The temperature of the saturated specimen was gradually increased from 25 to 50 °C. When the specimen temperature reaches 50 °C, the drain valve is opened to allow the specimen to drain and consolidate until the pore water pressure dissipates to a stable level.

At this point, then the drain valve is closed for non-drainage cooling test, saturated specimen temperature from 50 °C gradually reduced to 25 °C. Since no cooling equipment is installed in the pressure chamber, the entire cooling process mainly relies on natural cooling. An electric fan with constant power is used for auxiliary cooling, which not only accelerates the completion of the cooling process but also ensures that the cooling rate is approximately the same in each cooling test. The relative time for each 1 °C drop in temperature (with the time at 50 °C as the initial time), the pressure drain readings around the pressure chamber, and the pore water pressure readings were recorded as the temperature was gradually reduced from 50 to 25 °C. The whole cooling process took about 90 min. The entire cooling process takes about 90 min, then the drain valve is opened to allow the specimen to drain or absorb water, and the pore water pressure is dissipated.

2.2 Thermal Consolidation Processes Under Isothermal Conditions

2.2.1 Pore Water Pressure Dissipation Process

Figure 2.4 shows the dissipation process curves of pore water pressure when the peripheral pressure σ_3 is 50 kPa, 100 kPa, and 150 kPa, respectively. It can be seen that the dissipation curves of pore water pressure are very similar at different confining pressures. At the beginning of consolidation, the normalized pore water pressure u/σ_3 of most specimens is close to 1. Only at 150 kPa, $u/\sigma_3 = 0.95$, which is slightly lower. Throughout the consolidation process, the dissipation of pore water pressure mainly occurs in the first 200 min, and $u/\sigma_3 < 0.2$ at 200 min, and then the rate of pore water pressure dissipation gradually decreases.

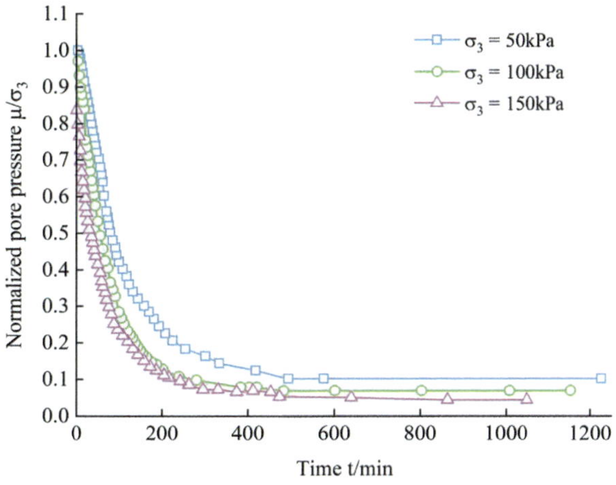

Fig. 2.4 Dissipation of pore pressure with time in primary consolidation processes

After about 1000 min of consolidation process, u/σ_3 dissipates to below 0.1 and stabilizes. Figure 2.4 shows that the larger the effective consolidation pressure σ_3 is, the smaller the normalized pore water pressure u/σ_3 is. However, there seems to be little difference in the pore water pressure values when the effective consolidation pressure σ_3 is 50 kPa, 100 kPa, and 150 kPa, at which time the pore pressure u is 5 kPa, 7 kPa, and 6 kPa, respectively.

2.2.2 *The Variation of Volumetric Strain*

Figure 2.5 gives the volume strain–time relationship curves for the specimen at 25 °C isothermal condition with a perimeter pressure $\sigma_3 = 150$ kPa. In the figure, curve 1 shows the volume strain ε_v versus time t measured from the drainage inside the specimen (the counterpressure control system), and curve 2 shows the volume strain ε_v versus time t measured from the volume change of water inside the pressure chamber (the perimeter pressure control system).

As can be seen from Fig. 2.5, the relationship between the body strain with time obtained by the two measurement methods at room temperature (25 °C) is basically the same. That is, because the soil particles and pore water in the specimen have no expansion or contraction effect, the volume change of the specimen is equal to the amount of pore water discharged from the specimen, which can be measured by the change in the volume of water inside the pressure chamber to measure the volume change of the specimen under room temperature conditions. In fact, the variation process of volumetric strain under confining pressures σ_3 of 50 and 100 kPa, respectively, is similar to the above-described case.

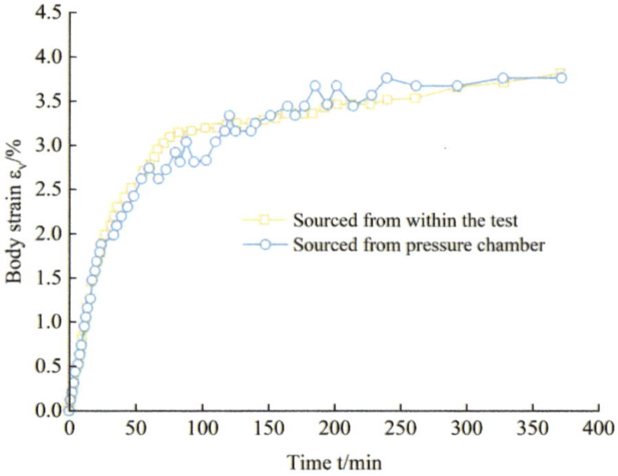

Fig. 2.5 Volumetric strain with time in consolidation processes under normal temperature

2.2.3 *Temperature Fluctuations During Thermal Consolidation*

As can be seen in Fig. 2.5, the strain of the solidified body measured by the change in water volume within the pressure chamber generally fluctuates in the later stages of consolidation. This is related to the contraction or expansion of the water volume within the pressure chamber caused by the dissipation of heat within the pressure chamber, i.e., the decrease or increase in temperature. In fact, the temperature value is measured by the temperature change of the water inside the pressure chamber, and the fluctuation range of the temperature inside the pressure chamber is related to the accuracy of the temperature control system. From the curve 1 in Fig. 2.6, it can be seen that the fluctuation trend of the water volume within the pressure chamber is consistent with the fluctuation trend of the temperature.

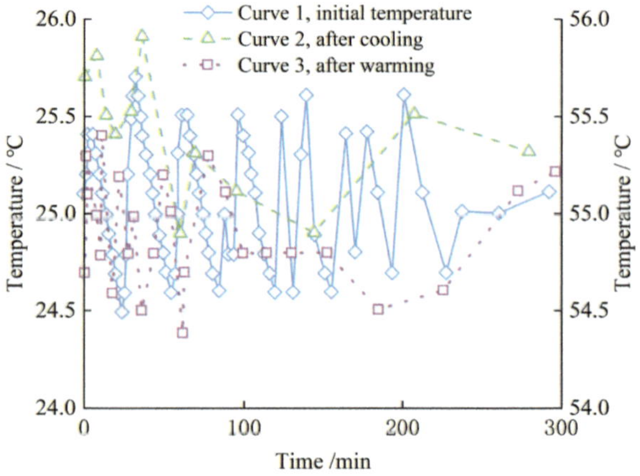

Fig. 2.6 Temperature fluctuations under isothermal condition

In fact, in the early stage of consolidation and drainage, due to the volume change of the specimen, i.e., the consolidation drainage is relatively large, so the temperature fluctuations on the change of the volume of water in the pressure chamber have a small effect, while in the late stage of consolidation, the drainage change is very small, the consolidation is stabilized, and the volume change of the water in the pressure chamber caused by temperature fluctuations is relatively large. As can be seen from curve 1 in Fig. 2.6, the temperature fluctuation range during isothermal consolidation is (25 ± 0.5) °C, and it is generally believed that this accuracy can meet the requirements of thermal consolidation test.

2.3 Thermal Consolidation Response Caused by Temperature Changes

2.3.1 Temperature Changes During Heating or Cooling Without Drainage

Figure 2.7 gives the change process of temperature from 25 °C to 50 °C when σ_3 is 50 kPa, 100 kPa, and 150 kPa, respectively, under undrained condition, as well as the change process of decreasing from 50 °C to 25 °C after the completion of consolidation under this temperature. It can be seen that the temperature change process under different enclosing pressures is almost the same, both in the warming process and the cooling process. In the initial stage of the warming process, the rate of temperature increase is large, and with the gradual increase of the sample temperature, the rate of temperature increase gradually slows down. The entire heating process lasted approximately 100 minutes. In the first 50 minutes, the specimen temperature rose from 25°C to 41°C, with a temperature increase of 16°C. In the subsequent 50 minutes, the temperature gradually approached a stable value of 50°C, with an additional increase of 9°C.

Figure 2.7 also shows that after about 90 min, the temperature can be reduced from 50 to 25 °C, and this process seems to be faster than the heating process. Moreover, during the whole cooling process, the difference between the temperature in the pressure chamber and the ambient temperature in the test chamber gradually decreases, so the rate of temperature reduction also decreases, and the temperature finally tends to 25 °C, a stable value.

2.3.2 Pore Water Pressure Evolution Process

Figure 2.8 gives the change process of pore water pressure during the whole process of undrained heating, isothermal drainage, undrained cooling, and isothermal water absorption of the specimen at the peripheral pressures of 50 kPa, 100 kPa, and

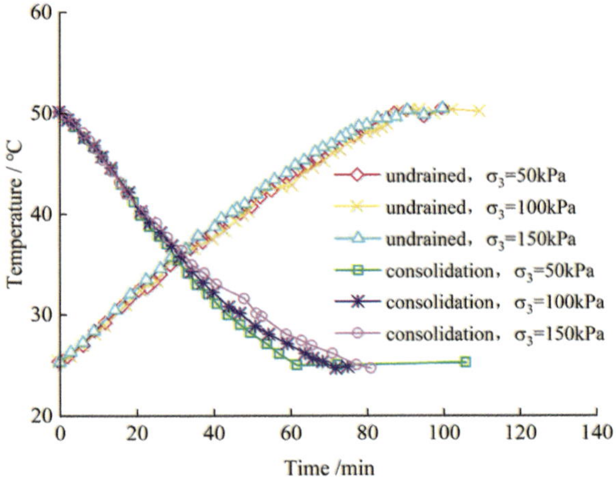

Fig. 2.7 Temperature evolutions with time by heating or cooling under undrained conditions

150 kPa, respectively. It can be seen that the evolution law of pore water pressure is similar under different enclosure pressures. In the process of gradually increasing temperature, the pore water pressure inside the saturated specimen caused by the temperature also increases. When the temperature reaches the set maximum temperature (50 °C), the pore water pressure inside the specimen also reaches the peak value. In addition, the peak pore water pressure appears to increase slightly with increasing ambient pressure. For example, the peak pore water pressure u was 34.2 kPa, 40.1 kPa, and 37.5 kPa when the perimeter pressure was 50 kPa, 100 kPa, and 150 kPa, respectively.

Figure 2.9 gives the peripheral pressure of 150 kPa, the specimen in the undrained warming and cooling process of the pore water pressure over time curve. It can be seen that with the increase in temperature, caused by the temperature of the pore water pressure increases, when the temperature rises from 25 to 50 °C, the pore

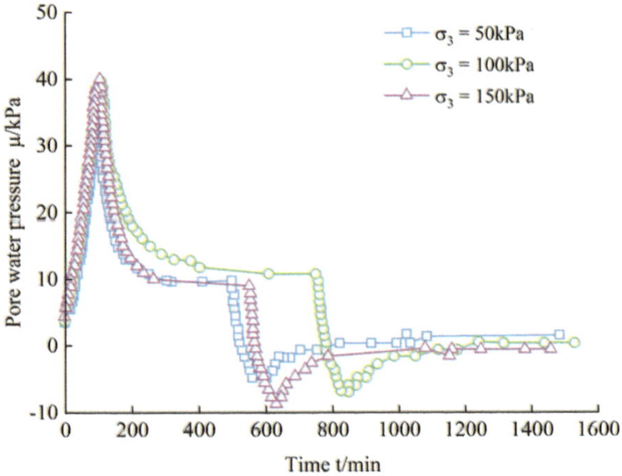

Fig. 2.8 Pore pressure evolutions under various confining pressures

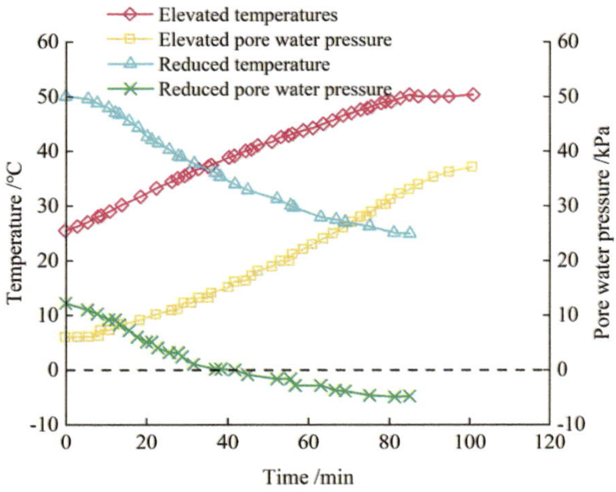

Fig. 2.9 Variation of pore pressure with temperature under undrained conditions

water pressure from 6 kPa rose to 37.5 kPa, and its rate of growth and the rate of temperature increase are basically the same, while the former seems to lag a little.

After the specimen temperature rises to 50 °C, the specimen begins to undergo drainage consolidation, and the pore water pressure gradually dissipates. When the consolidation time is about 200 min, the pore water pressure tends to a stable value, that is, there is a residual pore water pressure. When the peripheral pressure is 50 kPa, 100 kPa, and 150 kPa, the residual pore water pressure u is 10.9 kPa, 10.2 kPa, and 11.6 kPa, respectively.

As can be seen in Fig. 2.9, when the specimen temperature is reduced from 50 to 25 °C under the undrained condition, the pore water contraction deformation within the specimen is much larger than that of the soil particles, so the pore water pressure decreases and produces the negative pore water pressure at the later stage, which is manifested as an irreversible pore water pressure change process. In fact, the coefficient of thermal expansion of soil particles is generally 1.5×10^{-5} °C^{-1}, while the coefficient of thermal expansion of water is about 2×10^{-4} °C^{-1}. Moreover, with the increase of the perimeter pressure, the negative pore water pressure is also larger. For example, when the peripheral pressure was 50, 100, 150 kPa, the maximum negative pore water pressure were – 3.1, – 5.2, – 5.4 kPa. From Fig. 2.9, it can also be seen that the rate of temperature drop and pore water pressure drop rate is also roughly the same, while the pore water pressure drop seems to be lagging behind the process of temperature drop. When the temperature inside the specimen is reduced to 25 °C, open the drain valve, then due to the existence of negative pore water pressure, the specimen is water absorption process, and pore water pressure gradually rises and tends to 0.

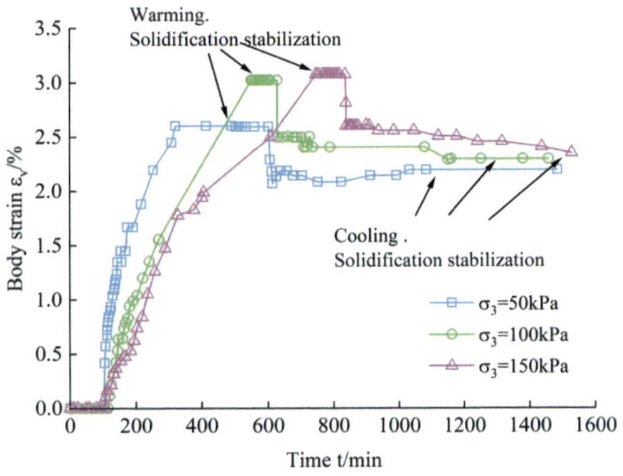

Fig. 2.10 Volumetric strain evolutions under various confining pressures

2.3.3 The Variation Law of Volumetric Strain

Figure 2.10 gives the variation process of the solidus strain ε_v with time t measured by the change of the pore water volume in the specimen during the pore water pressure evolution. It can be seen that the trend of the solidus strain of the specimen under different confining pressures is similar. Among them, the solidus strain of the specimen is zero at the stages of undrained warming (25–50 °C after about 400 min) and cooling (50–25 °C), while the specimen will be drained and absorbed during the two solidus processes of warming up to 50 °C and then cooling down to 25 °C, i.e., the volumetric strain of the specimen will decrease and increase, and the final solidus deformation tends to be stabilized.

In fact, as can be seen from Fig. 2.6, the temperature inside the pressure chamber (curve 2) at the stage of drainage consolidation after the temperature reaches 50 °C and the temperature inside the pressure chamber (curve 3) during the water absorption process after the temperature is further reduced to 25 °C will fluctuate to a certain extent, with the fluctuation ranges of 50 ± 0.5 and 25 ± 0.5 °C, which is able to satisfy the requirements of the isothermal consolidation test. The causes of temperature fluctuation can be attributed to heat loss in the pressure chamber caused by the difference between the laboratory ambient temperature and the temperature inside the pressure chamber, as well as fluctuations in temperature-controlled heating.

To further illustrate the dissipation characteristics of the pore water pressure after temperature increase and after further temperature reduction, Fig. 2.11 gives the process of pore water pressure dissipation and the change of water volume inside the specimen ($\sigma_3 = 150$ kPa) depicted with the beginning of consolidation drainage or water absorption as the starting point. From Fig. 2.11b, it can be seen that the volume of consolidation drainage after temperature increase is much larger than the volume of water absorption deformation after the temperature is lowered again. That is to say, after the process of temperature increase and decrease, the specimen overall produces obvious shrinkage deformation, which is manifested as an irreversible thermodynamic process. For example, when σ_3 was 50, 100, 150 kPa, respectively, the temperature increase to 50 °C after the solidus strain is generally 2.19 ~ 3.20%, while the temperature is then reduced to 25 °C after the solidus strain is generally − 0.1 ~ − 0.3%. From the comparison of Fig. 2.11a, b, it can be seen that the time required for the specimen solidus strain to reach the stabilized value seems to lag behind the time when the pore water pressure dissipation is completed.

Fig. 2.11 Comparison of consolidation processes after heating and cooling under drained conditions

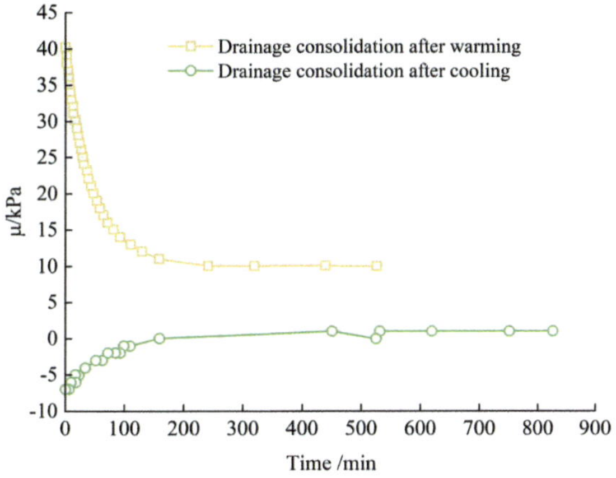

(a) Pore water pressure

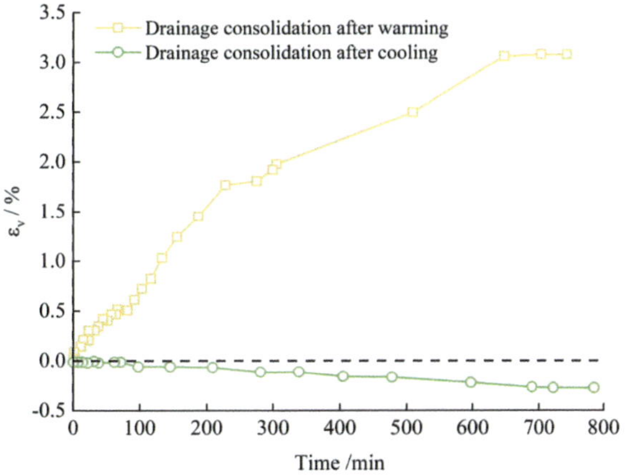

(b) Body strain

2.4 Summary

(1) At room temperature of 25 °C, the body strain measured in the pressure chamber is basically the same as that measured by the water displacement in the specimen, which indicates that the volume change of the specimen at room temperature can be determined by the change of the water volume in the pressure chamber at this time.

(2) The change trends of pore water pressure under different enclosing pressures are similar. In the process of gradually increasing temperature, the pore water pressure inside the saturated specimen caused by the temperature increases continuously. When the temperature reaches the set temperature, the pore water pressure inside the specimen also reaches the peak value. And the peak pore water pressure seems to increase slightly with the increase of the perimeter pressure.

(3) The trend of solidus strain change of the specimens under different peripheral pressures is similar. During the two drainage consolidation processes of 50 °C after temperature increase and 25 °C after further temperature reduction, the specimens were drained and absorbed water, respectively, i.e., the volumetric strain of the specimens would decrease and increase, and the final consolidation deformation tends to be stabilized.

(4) The amount of consolidation drainage after the temperature increase is much larger than the amount of water absorption deformation after the temperature decrease. That is to say, after the process of temperature increase and decrease, the specimen overall shows obvious shrinkage effect.

Chapter 3
Thermal Consolidation Behavior of Saturated Clay Induced by Multi-stage Heating and Cooling Processes

3.1 Experimental Overview

Using the self-developed testing equipment introduced in Chap. 2, experimental studies were conducted to investigate the thermal characteristics of saturated clay under the coupled effects of temperature and confining pressure. The testing apparatus includes a temperature control system, a confining pressure control system, an excess pore pressure control system, and a pore water pressure measurement system. The experimental design comprises two comparative groups: drained and undrained tests, with both groups having a temperature range of 20–80 °C and four consolidation pressures set at 50, 100, 150, and 200 kPa. Based on the experimental results, the evolution processes of temperature, pore water pressure, and consolidated strain during the thermal consolidation of the samples were analyzed. Additionally, the evolution of pore water pressure and consolidated strain was explained through an analysis of thermal characteristic mechanisms.

3.1.1 Preparation of Experimental Soil Samples

The soil used in the experimental study is a type of clay commonly found in China. The basic physical property parameters of the soil, measured through indoor experiments following the procedures and methods outlined in the "Standards for Geotechnical Testing Methods," are presented in Table 3.1.

(1) After air-drying, crushing, and passing through a 0.5 mm sieve, the field-collected soil samples were evenly spread in an iron tray and placed in an oven for drying for 8 h.
(2) Weigh 1500 g of the dried soil sample using an electronic balance and place it in a plastic basin. Use a spray bottle to evenly spray water onto the soil in the basin, taking special care not to let the water reach the bottom or sides of the

G. Yang and B. Bai, *Experimental and Theoretical Study on Thermal Consolidation Behavior of Saturated Clay*, https://doi.org/10.1007/978-981-96-6909-7_3

Table 3.1 Basic physical parameters of the clay used in the experiments

Specific gravity	Dry unit weight (kN/m^3)	Saturated moisture content (%)	Saturation degree (%)	Liquid limit (%)	Plastic limit (%)	Plasticity index (%)
2.47	11.97	44.2	97.84	71.8	39.2	32.6

basin, as this would cause uneven moisture distribution. Approximately 20 mL of water is sprayed into the basin each time, followed by thorough mixing of the soil before the next spraying. This step is repeated three times until the water content of the soil sample reaches the desired level of 30%.

(3) The thoroughly mixed wet soil is placed into a plastic sealed bag, excess air is expelled, and the bag is sealed for storage. The soil is allowed to fully saturate for a period of one day.
(4) After the wet soil has been fully saturated, samples are taken from each sealed bag to determine the moisture content. Three sets of soil samples, each weighing 15–30 g, are collected from different locations within each sealed bag to ensure that the measured moisture content is accurate and representative.
(5) The moisture content determined in step (4) is 29.7%, which differs from the target moisture content of 30% by no more than 1%, allowing for the continuation to the next step of sample preparation.
(6) Take 152 g of soaked wet soil and prepare a soil sample using the layered compaction method. Evenly divide the weighed wet soil into five layers and pour them into a compactor equipped with three valves. Each layer should be compacted 15 times until it reaches a height of 8 mm and a diameter of 3.91 mm. This will result in a soil sample with a density of 1.58 g/m^3 and a dry density of 1.22 g/m^3.
(7) Place the prepared sample into a saturator for vacuum pumping and saturation with distilled water. After saturation, the measured moisture content is 44.2% and the saturation degree is 97.8%.

3.1.2 Experimental Plan

The ambient temperature during the experiment is approximately 20 °C. The confining pressures used in the tests are 50, 100, 150, and 200 kPa. The specific plan and corresponding identifiers are shown in Table 3.2.

Plan 1 is a drained test: For samples A1, A2, A3, and A4, confining pressures of 50 kPa, 100 kPa, 150 kPa, and 200 kPa are applied, respectively. The four soil samples undergo the same procedures except for the different confining pressures applied.

Heating (Cooling) Process: The heating process from 20 to 40 °C is described as follows: after applying confining pressure, the test specimen is allowed to consolidate through drainage until the pore water pressure has fully dissipated. Subsequently, the

Table 3.2 Experimental plan

Plan	Sample ID	Confining pressure (kPa)	Initial temperature (°C)	Maximum temperature (°C)	Final temperature (°C)	Description of experimental conditions
1	A1	50	20	80	20	After the initial drained consolidation is completed, the sample undergoes drained consolidation after each increase (or decrease) in temperature
	A2	100	20	80	20	
	A3	150	20	80	20	
	A4	200	20	80	20	
2	B1	50	20	80	20	After the initial drained consolidation is completed, the sample remains undrained after each increase (or decrease) in temperature
	B2	100	20	80	20	
	B3	150	20	80	20	
	B4	200	20	80	20	

temperature control device is adjusted to set the upper limit of heating at 40 °C. The pore water drainage valve is closed, and timing begins simultaneously. Data points are recorded for time, temperature, pore water pressure, and changes in water levels in both the drainage pipe of the pressure chamber and the pore water drainage pipe. The entire heating (or cooling) process lasts for 120 min.

Drainage Consolidation Process: After the aforementioned 120 min heating (or cooling) process, the indicators within the soil sample have reached a stable state. The pore water drainage valve is then opened to allow the drainage of pore water from the soil sample. Simultaneously, timing begins, and data points are recorded for time, temperature, pore water pressure, and changes in water levels in both the drainage pipe of the pressure chamber and the pore water drainage pipe. This process lasts for 1320 min.

Each temperature variation stage encompasses the two processes described above: the undrained heating (or cooling) process followed by the drainage consolidation process. These two processes are conducted consecutively without interruption, with a single temperature variation stage lasting 1440 min. The temperature variation stages include: the heating stage from 20 to 40 °C, the heating stage from 40 to 50 °C, the heating stage from 50 to 60 °C, the heating stage from 60 to 70 °C, the heating stage from 70 to 80 °C, the cooling stage from 80 to 70 °C, the cooling stage from 70 to 60 °C, the cooling stage from 60 to 50 °C, the cooling stage from 50 to 40 °C, and the cooling stage from 40 to 20 °C. These stages are carried out in the specified order without any time intervals between them. The temperature gradient for the first heating process is set at 20 °C, primarily because the difference between 30 °C and room temperature is minimal, allowing direct heating to 40 °C to facilitate the study of the effects of higher temperatures on the consolidation tests. Similarly, the final cooling gradient is also set at 20 °C, enabling direct comparison with the changes observed in the first stage.

Scheme 2 is an undrained test: confining pressures of 50 kPa, 100 kPa, 150 kPa, and 200 kPa are applied to samples B1, B2, B3, and B4, respectively. The four soil samples undergo the same process except for the application of different confining pressures.

Heating (Cooling) Process: Same as the heating (cooling) process in scheme 1. After a 120 min heating (cooling) process, it enters the isothermal non-draining stage, where the pore water drainage valve is not opened and the pore water in the soil sample cannot be drained. This process lasts for 1320 min. Each temperature change stage includes the two processes mentioned above. Firstly, an undrained heating (cooling) process is carried out, followed by an undrained process. The two processes are continuous and uninterrupted, with one temperature change stage lasting 1440 min. The sample temperature change stage includes: 20–40 °C heating stage, 40–50 °C heating stage, 50–60 °C heating stage, 60–70 °C heating stage, 70–80 °C heating stage, 80–90 °C cooling stage, 70–60 °C cooling stage, 60–50 °C cooling stage, 50–40 °C cooling stage, and 40–20 °C cooling stage. Each temperature change stage is carried out continuously in the above order without any time interval.

3.2 Analysis of Initial Consolidation Test Results

(1) Analysis of Changes in Pore Water Pressure During Initial Consolidation

Figure 3.1 presents the dissipation curves of pore water pressure during initial drainage consolidation of soil samples at an initial temperature of 20 °C, under confining pressures (σ_3) of 50, 100, 150, and 200 kPa. It can be observed that the dissipation patterns of pore water pressure are generally consistent across different confining pressures. At the onset of drainage consolidation, the ratio of pore water pressure to confining pressure (normalized pore water pressure) was recorded as 0.94, 0.93, 0.94, and 0.97 for confining pressures of 50 kPa, 100 kPa, 150 kPa, and 200 kPa, respectively (subsequent data, unless otherwise specified, will follow this order). During the first 20 min of drainage, the normalized pore water pressure decreased rapidly and exhibited a near-linear trend; subsequently, the rate of change slowed down. At 20 min, the normalized pore water pressures for the four confining pressures were reduced to 0.16, 0.26, 0.38, and 0.59, respectively. The majority of pore water pressure dissipation occurred within the first 200 min, with normalized pore water pressure dropping below 0.1 by the 200 min mark. From 200 min to the end of the test at 1320 min, normalized pore water pressure fluctuated around a stable value. At stability, the normalized pore water pressures for the four confining

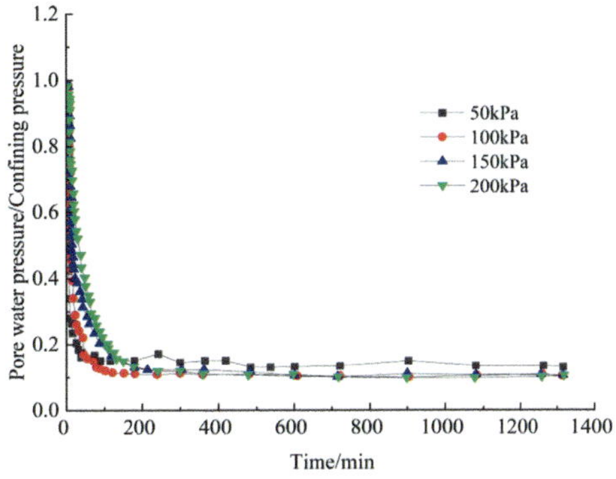

Fig. 3.1 Changes in normalized pore water pressure during the initial drainage consolidation process

pressures were measured as 0.04, 0.02, 0.01, and 0.01, respectively. This data indicates that as the effective confining pressure increases, the normalized pore water pressure decreases. However, the stable values of pore water pressure show minimal variation among the different confining pressures, all being approximately 2 kPa at confining pressures of 50, 100, 150, and 200 kPa. Within the first 120 min, the dissipation patterns of pore water pressure under different confining pressures were largely consistent, albeit with varying rates of dissipation. The curve for the 50 kPa confining pressure exhibited the fastest rate of change and reached stability first, whereas the curve for the 200 kPa confining pressure displayed the slowest change and took the longest to stabilize. This suggests that lower confining pressures result in quicker dissipation of pore water pressure.

(2) Analysis of Initial Consolidation Strain

Figure 3.2 illustrates the variation of volumetric strain with time during the isothermal drainage consolidation process at a temperature of 20 °C for soil samples subjected to confining pressures of 50, 100, 150, and 200 kPa. It is assumed that under normal temperature conditions, there are no expansion or contraction effects of soil particles and pore water within the samples; thus, the change in volume of the soil sample is represented by the volume of pore water expelled. The analysis shows that during the isothermal drainage consolidation process at 20 °C, the trends of volumetric strain remain consistent under different confining pressures. In the first 10 min of drainage consolidation, the change in volumetric strain occurs rapidly, and the curve exhibits an almost linear pattern. From 10 to 120 min, the rate of change in volumetric strain decreases and gradually approaches stability. By 200 min, the volumetric strain stabilizes and remains constant until the end of the consolidation phase. At the conclusion of the drainage consolidation process, the volumetric strains for the soil samples at confining pressures of 50 kPa, 100 kPa, 150 kPa, and 200 kPa were recorded as 9.79%, 10.94%, 12.08%, and 15.42%, respectively. This indicates that higher confining pressures result in larger volumes of pore water expelled during consolidation, leading to increased volumetric strain.

Moreover, the time required for the volumetric strain to stabilize varies with confining pressure. The stabilization times for volumetric strain under confining

Fig. 3.2 Changes in initial consolidation volumetric strain

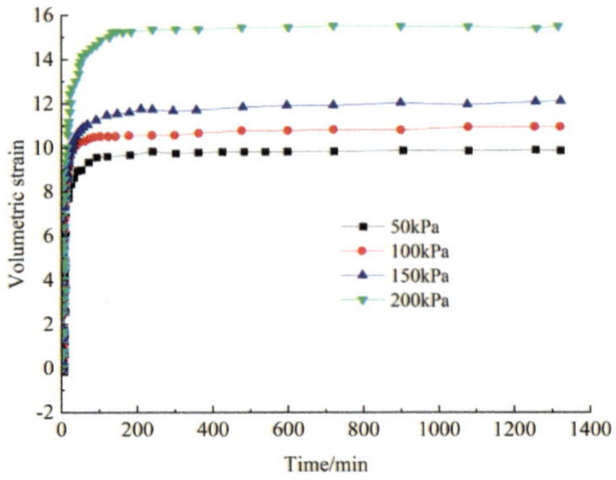

pressures of 50 kPa, 100 kPa, 150 kPa, and 200 kPa were approximately 40 min, 60 min, 90 min, and 110 min, respectively.

Comparative analysis of Figs. 3.1 and 3.2 reveals that as pore water pressure continuously dissipates, the volumetric strain measured by the expelled drainage volume of the soil sample steadily develops. When the pore water pressure dissipation reaches a stable state, the volumetric strain also approaches stability. The stabilization time for volumetric strain aligns closely with the completion time of pore water pressure dissipation.

During the expulsion of pore water, the soil particles within the sample undergo rearrangement due to the interplay of temperature and pressure, accompanied by changes in volume.

3.3 The Thermal Consolidation Process of Multi-stage Heating/Cooling Under Drainage Condition

3.3.1 Temperature Variation Process

Figure 3.3 presents the temperature–time variation process of samples during the multi-stage heating test with drainage, where the temperature increases from 40 to 50 °C. It can be observed that the temperature changes of the four samples under different confining pressures are largely consistent. After reaching the set value (i.e., 50 °C), the temperature stabilizes, with fluctuations within ± 0.5 °C. Similarly, the temperature variation process during the cooling phase under different confining pressures (e.g., cooling from 70 to 60 °C) is also quite consistent. This indicates that the confining pressure has little impact on the temperature variation process of the samples.

Figure 3.4 shows the temperature change process of the sample under different heating levels (confining pressure of 150 kPa). For ease of comparison, the temperature changes ΔT at different heating stages are placed in the same coordinate system,

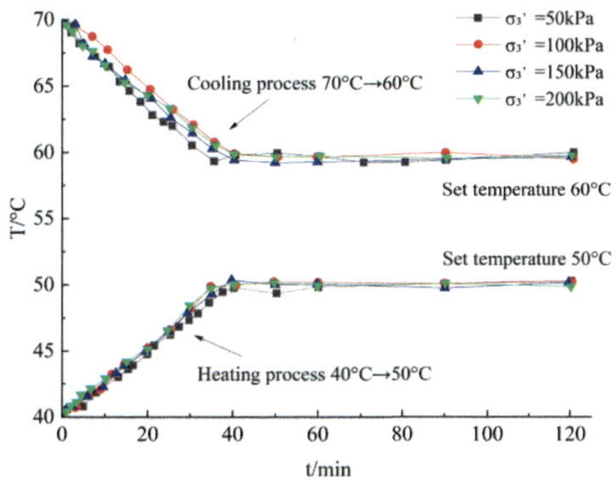

Fig. 3.3 Heating or cooling process of samples under different confining pressures

represented in increments of temperature. It can be seen that the temperature rise process at each temperature level is similar and almost linearly increases to reach the set value. Of course, when the temperature level is low (such as 40 → 50 °C), the temperature rise process is faster (about 40 min). As the temperature level increases (70 → 80 °C), the rate of temperature rise slows down significantly (about 110 min), which is related to the heat loss of the sample.

After completing the undrained heating and isothermal drainage consolidation process step by step, the samples are subjected to undrained cooling and isothermal drainage tests step by step. The process of cooling the sample from 70 to 60 °C (i.e., 70 → 60 °C) is also shown in Fig. 3.3. It can be seen that the process of undrained cooling under different confining pressures is basically the same, indicating that the influence of confining pressure on the temperature change of the sample can be ignored.

As shown in Fig. 3.4, in different cooling stages, the slope of the cooling curve from 80 to 70 °C is the largest, and it drops quickly to the set temperature of 70 °C, taking 25 min. As the cooling level decreases, the temperature drop gradually slows down (such as 50 → 40 °C), and the required time is 60 min, indicating that the

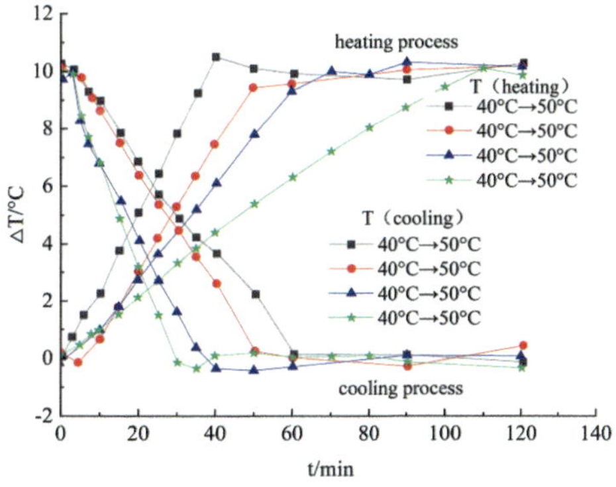

Fig. 3.4 Heating or cooling process of samples at different temperature levels

smaller the difference with the ambient temperature, the slower the temperature drop.

3.3.2 Pore Water Pressure Variation Process

Figure 3.5 illustrates the pore water pressure increase process at different heating levels, as well as the evolution of the pore water pressure to confining pressure ratio (u/σ_3) over time during the drainage process. During the undrained heating process at the same temperature level, the time required for the temperature to reach the set value is roughly the same under different confining pressures, and the time taken for u/σ_3 to peak is also quite similar. Furthermore, at any given temperature level, a lower confining pressure corresponds to a steeper slope in the u/σ_3 curve during the heating phase, and the peak value of u/σ_3 gradually increases, approaching the effective consolidation pressure. This indicates that under the coupled effects of temperature and confining pressure, a higher confining pressure results in weaker variations in pore water pressure due to temperature changes, which is primarily related to the greater amount of consolidation drainage that has already occurred during the initial consolidation process at higher confining pressures. After reaching the peak value of u/σ_3, the pore water pressure continues to fluctuate with temperature variations.

From Fig. 3.5a, it is evident that when the temperature increases from 20 to 40 °C (with a temperature increment of $\Delta T = 20$ °C), the rise in pore water pressure is significant. Comparing Fig. 3.5b–e, it can be seen that as the temperature levels increase, the difference in expansion between soil particles and pore water caused by the temperature also increases, leading to a trend of increasing pore water pressure. For instance, when $\sigma_3 = 50$ kPa, during the temperature increments of 40 °C → 50 °C, 50 °C → 60 °C, 60 °C → 70 °C, and 70 °C → 80 °C (each with a temperature increment of $\Delta T = 10$ °C), the corresponding pore water pressure ratios u/σ_3 are 0.24, 0.26, 0.32, and 0.36, respectively.

Figure 3.6 illustrates the variation of pore water pressure and the normalized pore water pressure ratio (u/σ_3) over time at different confining pressures during each cooling stage. It can be observed that during the undrained cooling process (the first 120 min), the normalized pore water pressure (u/σ_3) continuously decreases and exhibits negative values. In fact, during the cooling process, both the soil particles and water within the sample contract, with their respective volumetric thermal expansion coefficients being approximately 4.5×10^{-5} °C^{-1} for soil particles and 6.0×10^{-4} °C^{-1} for water. This indicates that the contraction of water is significantly greater than that of the soil particles, resulting in a difference that even leads to negative pore water pressures.

Once the temperature stabilizes after the cooling, the pore water pressure decreases to its minimum value. It is evident that the maximum value of negative pore water pressure diminishes as the temperature level decreases. For example, under a confining pressure of $\sigma_3 = 50$ kPa, the temperature changes from 80 °C to 70 °C,

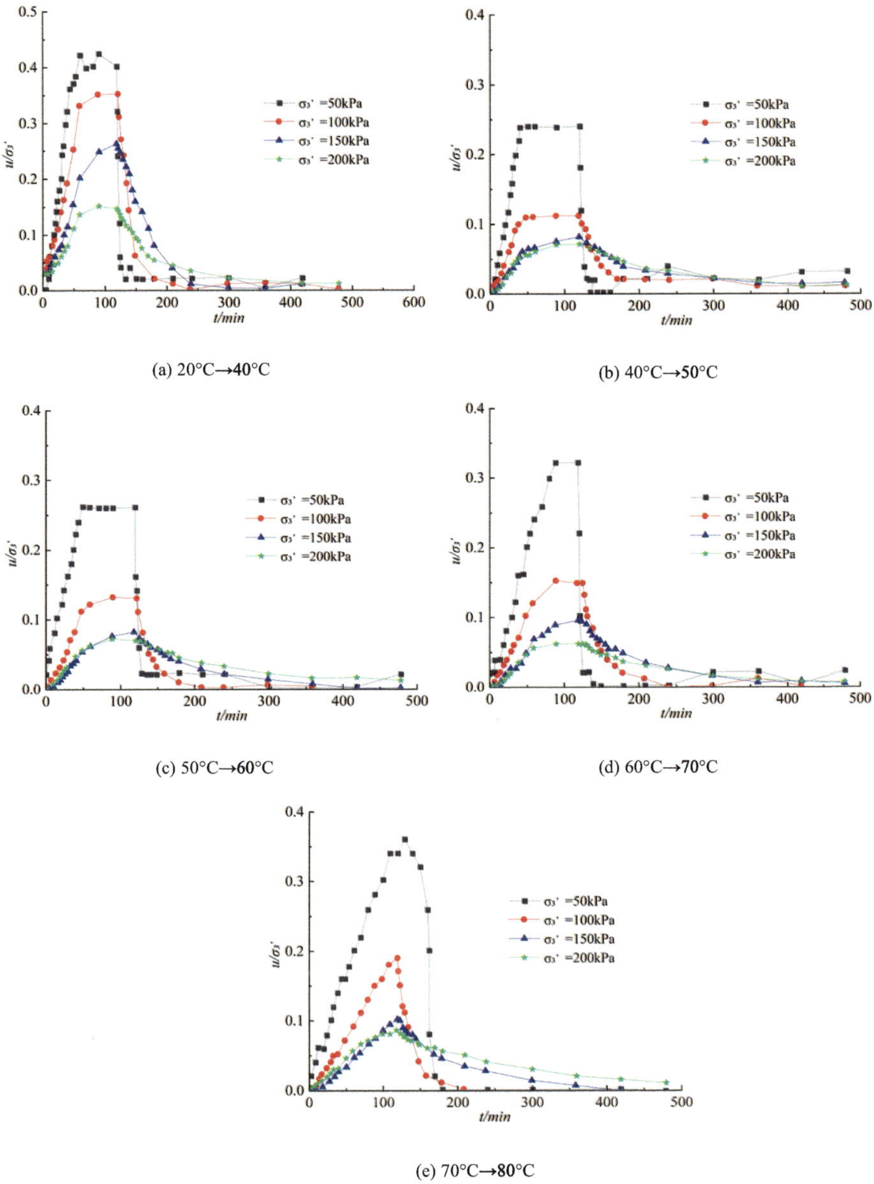

(a) 20°C→**40**°C

(b) 40°C→**50**°C

(c) 50°C→**60**°C

(d) 60°C→**70**°C

(e) 70°C→**80**°C

Fig. 3.5 Development and dissipation of pore pressure induced by heating

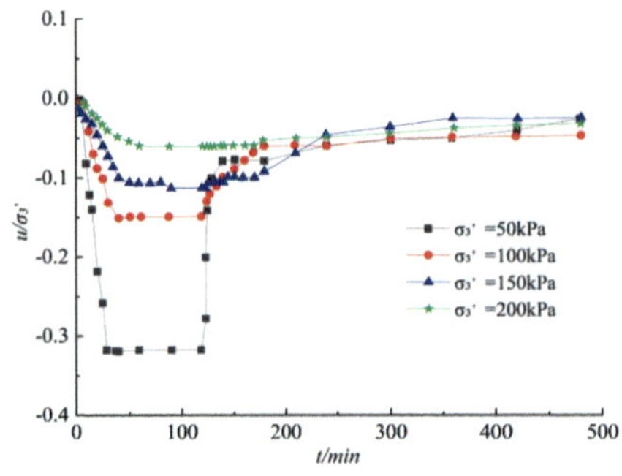

(a) 80°C→**70**°C

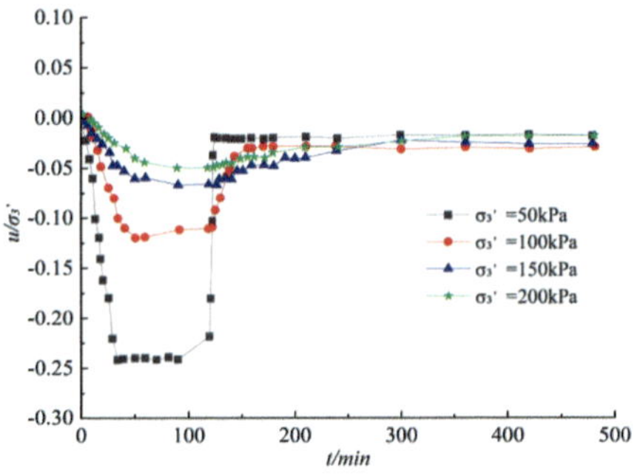

(b) 70°C→**60**°C

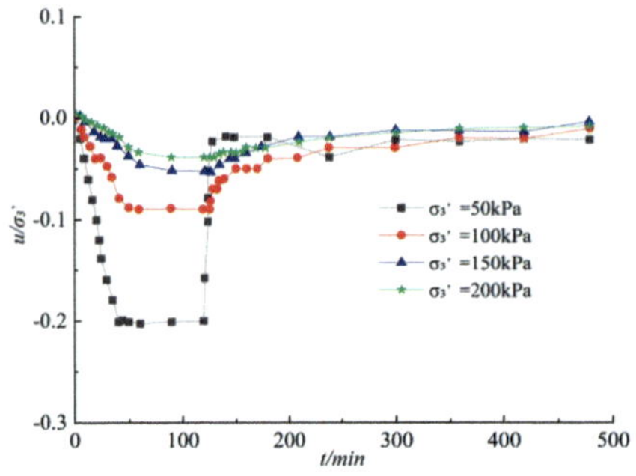

(c) 60°C→**50**°C

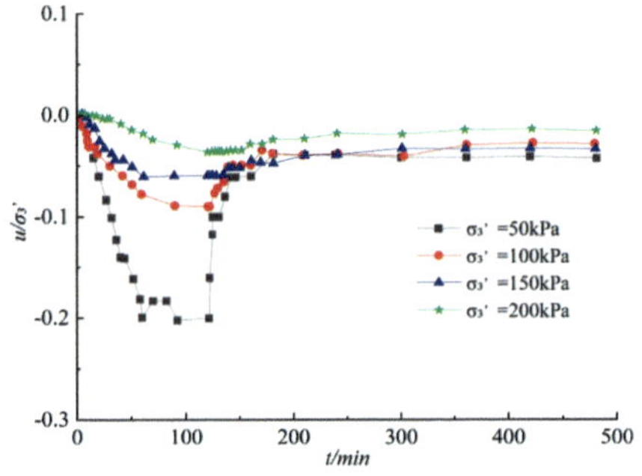

(d) 50°C→**40**°C

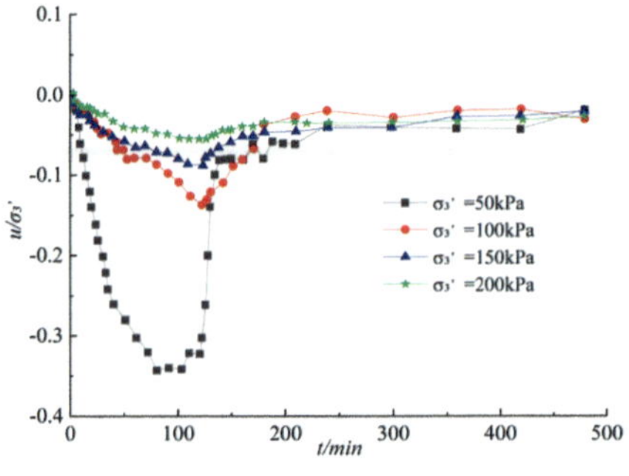

(e) 40°C→**20**°C

Fig. 3.6 Development and dissipation of pore pressure induced by cooling

70 °C to 60 °C, 60 °C to 50 °C, and 50 °C to 40 °C (with a temperature decrease of $\Delta T = 10$ °C), the corresponding u/σ_3 values are – 0.32, – 0.24, – 0.20, and – 0.20, respectively.

During the consolidation process following the temperature drop (i.e., after 120 min), due to the existence of negative pore water pressure, the sample absorbs water, causing the pore water pressure to gradually rise toward zero while exhibiting some fluctuations. Furthermore, at the same temperature level, the minimum value of u/σ_3 during the cooling stage decreases with increasing confining pressure, indicating that under the coupled effects of temperature and confining pressure, the influence of temperature on pore water pressure weakens as the confining pressure increases.

3.3.3 The Process of Volumetric Strain Variation

As shown in Fig. 3.7, after each stage of temperature increase and subsequent drainage, the consolidated volumetric strain increases with the rise in temperature levels. For example, when $\sigma_3 = 50$ kPa, the volumetric strains at the end of drainage for temperature levels of 20–40 °C, 40–50 °C, 50–60 °C, 60–70 °C, and 70–80 °C are 1.12%, 1.94%, 2.87%, 4.09%, and 5.86%, respectively. In contrast, during the staged cooling process (see Fig. 3.8), the consolidated volumetric strain represents a continuous decrease (as the soil sample absorbs water). For instance, when $\sigma_3 =$ 50 kPa, the volumetric strains at the end of each cooling stage from 80 °C to 70 °C, 70 °C to 60 °C, 60 °C to 50 °C, 50 °C to 40 °C, and 40 °C to 20 °C are 5.12%, 4.38%, 3.82%, 3.46%, and 2.71%, respectively.

It can be observed that, whether during heating or cooling, the change in consolidated volumetric strain is greater at higher temperature levels. Moreover, the positive volumetric strain generated during the heating phase is larger than the negative volumetric strain produced during the cooling phase, indicating that the volume of water drained after heating is greater than the volume absorbed during the subsequent cooling. At the end of the entire staged heating and then staged cooling process, the overall effect relative to the initial volume of the sample shows a decrease, representing an irreversible volume change process. For example, when $\sigma_3 = 50$ kPa, this irreversible volumetric strain can reach $\varepsilon_v = 2.71\%$ (see Fig. 3.8b).

Table 3.3 further presents the cumulative volumetric strain of the sample at a specific temperature level during the gradual heating and cooling processes. It can be seen that, at the same temperature level, the amount of overall volumetric strain change decreases with increasing confining pressure. This is because, during the initial consolidation process, the initial deformation under higher confining pressure is significantly greater, resulting in a relative reduction in the volumetric strain caused by subsequent thermal loading. The final effect indicates a trend toward consistency in volumetric strain under the coupled effects of temperature and confining pressure. When the temperature is gradually increased from 20 °C to 80 °C, the cumulative volumetric strains ε_v corresponding to increasing confining pressures (σ_3 from 50 kPa to 100 kPa to 150 kPa to 200 kPa) are 5.86%, 5.42%, 4.95%, and 4.45%, respectively. After subsequently cooling back to 20 °C, the final cumulative volumetric strains ε_v are 2.71, 2.65, 2.39, and 2.26%.

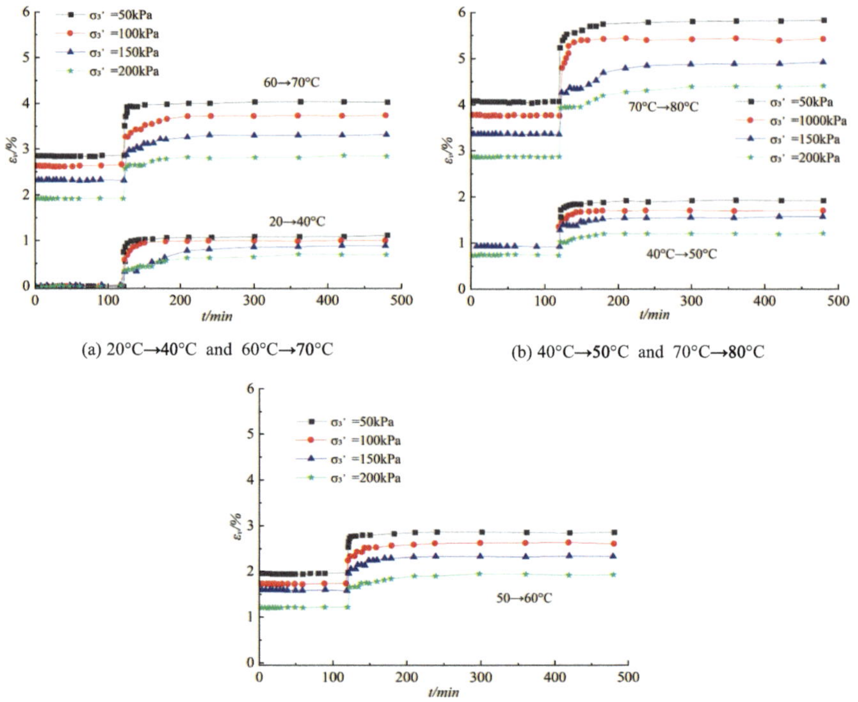

(a) 20°C→40°C and 60°C→70°C

(b) 40°C→50°C and 70°C→80°C

(c) 50°C→60°C

Fig. 3.7 Volumetric strain induced by heating

3.4 The Thermal Consolidation Process of Multi-stage Heating/Cooling Under Undrained Condition

Experiments indicate that under conditions of multi-stage heating and cooling without drainage between stages, the processes of temperature increase and decrease in the samples do not show significant differences from the evolution patterns under conditions of allowed drainage, and there is also no obvious relationship with the applied confining pressure.

From Table 3.3 and Fig. 3.9, it can be seen that during each heating stage (i.e., 20 → 40 °C, 40 → 50 °C, 50 → 60 °C, 60 → 70 °C, 70 → 80 °C), the variation pattern of the pore water pressure to confining pressure ratio u/σ_3 under different confining pressures is roughly the same. Specifically, when the temperature reaches the set value, u/σ_3 gradually reaches its maximum value, followed by a slight increase during the heating phase (20 → 40 °C) and a subsequent drop in the later stages (from 40 → 50 °C and beyond), ultimately entering a stable equilibrium state. When the confining pressure σ_3 is 50 kPa, during the heating stage from 60 to 70 °C, the maximum value of u/σ_3 can reach 1. With further temperature increases (from 70 to 80 °C), the pore water pressure does not increase further, indicating that the pore water pressure will

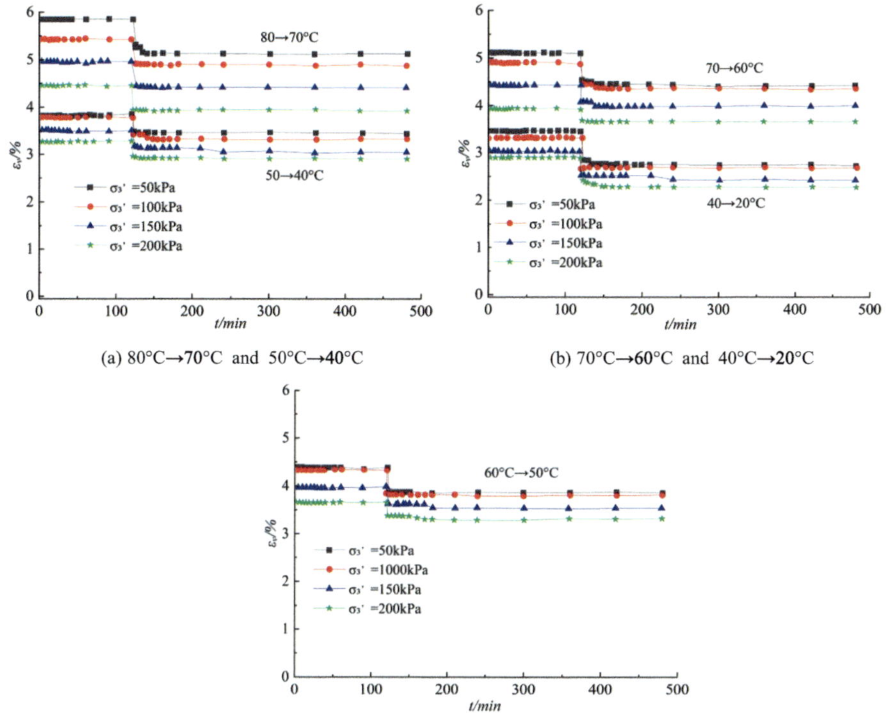

(a) 80°C→70°C and 50°C→40°C

(b) 70°C→60°C and 40°C→20°C

(c) 60°C→50°C

Fig. 3.8 Volumetric strain induced by cooling

Table 3.3 Accumulated volumetric strain induced by progressively heating and cooling

σ_3 (kPa)	Temperature variation	Cumulative volume strain				
		20 ↔ 40 °C	40 ↔ 50 °C	50 ↔ 60 °C	60 ↔ 70 °C	70 ↔ 80 °C
50	Heating	1.12	1.94	2.87	4.09	5.86
	Cooling	2.71	3.46	3.82	4.38	5.12
100	Heating	0.99	1.72	2.64	3.77	5.42
	Cooling	2.65	3.31	3.78	4.34	4.89
150	Heating	0.93	1.58	2.34	3.37	4.95
	Cooling	2.39	3.03	3.49	3.96	4.42
200	Heating	0.74	1.21	1.95	2.87	4.45
	Cooling	2.26	2.90	3.27	3.65	3.92

not exceed the applied confining pressure (as shown in Table 3.4). At this point, the strength of the sample has essentially been completely lost. Research shows that as the confining pressure increases (e.g., $\sigma_3 = 200$ kPa), the generated pore water pressure relative to the confining pressure decreases (i.e., u/σ_3 decreases), indicating that under the coupled effects of confining pressure and temperature, temperature changes lead to reductions in pore water pressure. In fact, when the temperature reaches its maximum value, the pore water pressure also reaches its peak, at which point the expansion of soil particles and water within the sample no longer progresses. Due to the creep and rearrangement of soil particles, at lower temperature stages, there is a trend of reduced pore volume between soil particles, while at higher temperature stages, there is a tendency for increased pore volume. Therefore, the pore water pressure will undergo slight increases or decreases before eventually stabilizing.

Figure 3.10 presents the changes in the pore water pressure to confining pressure ratio (u/σ_3) over time during the cooling stages at confining pressures σ_3 of 50, 100, 150, and 200 kPa. It indicates that during the undrained cooling phase, the pore water pressure continuously decreases. Once the temperature drops to the set value, the pore water pressure also reaches a minimum and then exhibits a certain degree of recovery before stabilizing. In fact, when the temperature has decreased to the

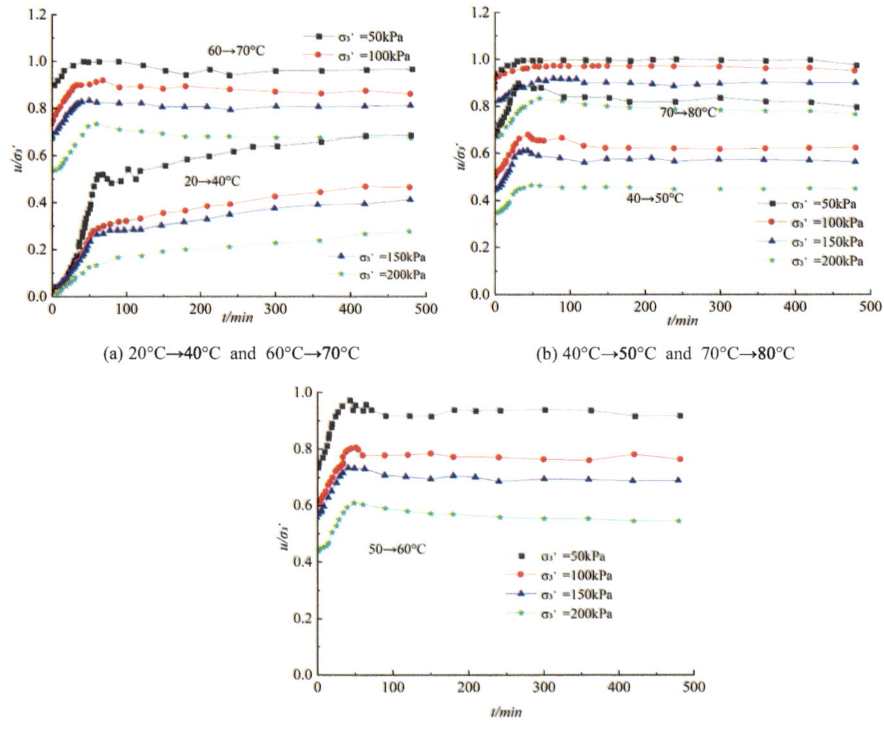

(a) 20°C→40°C and 60°C→70°C

(b) 40°C→50°C and 70°C→80°C

(c) 50°C→60°C

Fig. 3.9 Temporal variations of pore pressure induced by undrained heating

Table 3.4 Accumulated pore pressure induced by progressively heating and cooling

σ_3 (kPa)	Temperature variation	u/σ_3				
		20 ↔ 40 °C	40 ↔ 50 °C	50 ↔ 60 °C	60 ↔ 70 °C	70 ↔ 80 °C
50	Heating	0.72	0.88	0.98	1.00	1.00
	Cooling	0.18	0.46	0.58	0.68	0.84
100	Heating	0.51	0.67	0.81	0.92	0.97
	Cooling	0.24	0.43	0.54	0.65	0.74
150	Heating	0.45	0.61	0.73	0.83	0.91
	Cooling	0.19	0.41	0.51	0.63	0.74
200	Heating	0.35	0.46	0.61	0.74	0.84
	Cooling	0.16	0.28	0.36	0.44	0.54

set value, the contraction of the soil particles and water within the sample ceases, and the soil particles rearrange, which may result in a slight increase in pore water pressure before entering a new stable state (as seen in transitions from 80 to 70 °C and from 60 to 50 °C, shown in Fig. 3.10a, c).

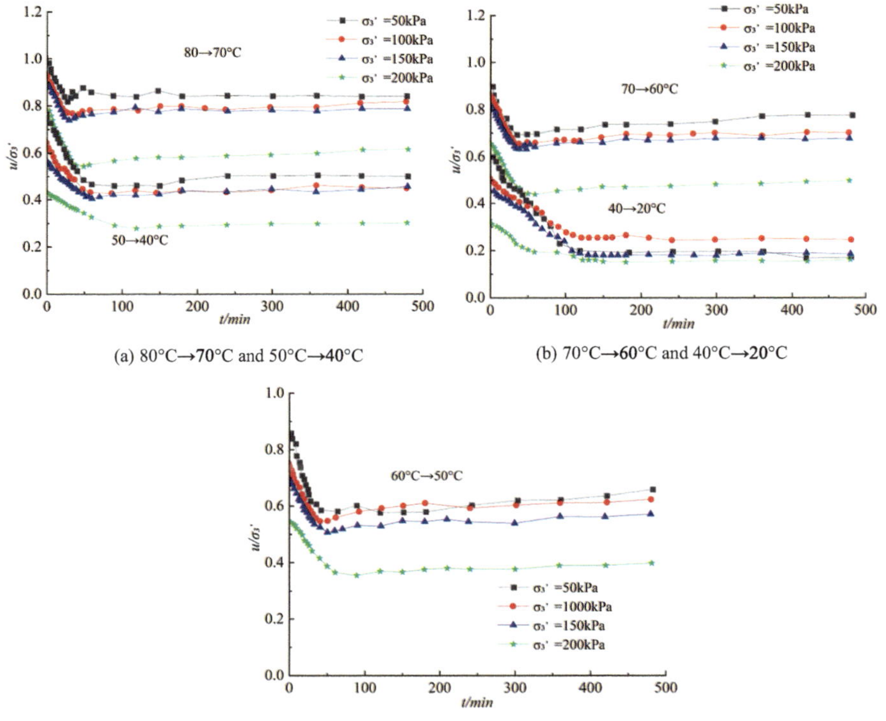

(a) 80°C→70°C and 50°C→40°C

(b) 70°C→60°C and 40°C→20°C

(c) 60°C→50°C

Fig. 3.10 Temporal variations of pore pressure induced by undrained cooling

It is also noteworthy that during lower temperature cooling stages, after the temperature drops to the set point, the pore water pressure continues to decrease slightly (such as from 40 to 20 °C, as depicted in Fig. 3.10b), indicating that the adjustment of soil particles leads to an increase in the pore volume of the soil mass.

Research shows that under the coupled effects of temperature and confining pressure, the greater the confining pressure, the smaller the ratio of pore water pressure to confining pressure, which indicates a reduction in pore water pressure due to temperature changes. For example, in the final cooling stage from 40 °C to 20 °C, the stabilized pore water pressures at confining pressures of 50 kPa, 100 kPa, 150 kPa, and 200 kPa are 9.0 kPa, 24.3 kPa, 28.5 kPa, and 32.4 kPa, respectively. This also illustrates that after a series of heating and cooling cycles, irreversible thermodynamic processes have occurred within the sample.

3.5 Summary

(1) The time required for the temperature to reach the set value is approximately the same under different confining pressures, and the time needed for the pore water pressure to reach its peak is also quite similar. As the confining pressure decreases, the slope of the pore water pressure increase becomes steeper, and the ratio of pore water pressure to confining pressure also increases, indicating that the influence of temperature effects under the coupled effect of temperature and confining pressure is greater at this point.

(2) In the undrained cooling process, the pore water pressure continuously decreases and exhibits negative pore water pressure. Once the temperature stabilizes at a lower value, the pore water pressure reaches its minimum. The influence of the temperature effect on pore water pressure diminishes as the confining pressure increases.

(3) The positive volumetric strain produced during the heating phase is greater than the negative volumetric strain generated during the cooling phase, indicating that the consolidation drainage volume after the temperature increase is greater than the water absorption volume after the temperature decrease. Compared to the initial volume of the sample, the overall effect at the end of the entire stepwise heating and subsequent stepwise cooling process manifests as a shrinkage of the sample's volume, representing an irreversible volume change process.

(4) In the undrained heating process with multi-stage temperature increments, the pore water pressure continuously increases, but its maximum value does not exceed the applied confining pressure. Furthermore, under the coupling effect of confining pressure and temperature, the influence of the temperature effect on pore water pressure diminishes as the confining pressure increases.

(5) During the undrained heating process with multi-stage temperature increments, when the pore water pressure reaches its maximum value, the creep of soil particles can lead to a trend of either decreasing or increasing pore volume between the soil particles, resulting in a slight increase or decrease in pore

water pressure. Additionally, when the temperature is reduced to the set value, the pore water pressure often exhibits a certain degree of rebound after dropping to a certain level, which is also caused by the rearrangement of soil particles.

(6) During the undrained heating process with multi-stage temperature increments, when the pore water pressure increases with rising temperature and remains elevated after cooling down to the set temperature, it indicates that the sample still exhibits an irreversible thermodynamic response due to the undrained heating–cooling cycle.

Chapter 4
Thermal Consolidation Characteristics of Saturated Clay Under Cyclic Temperature Loading

The thermal consolidation response of saturated clay under different drainage conditions and temperature paths under hot and cold cyclic temperature loading was investigated using self-developed test equipment. For example, the temperature path is 25–75–25 °C, the number of cycles is 6 times, the peripheral pressure is 50 kPa, 100 kPa, 150 kPa, and the pore water in the specimen is always in the state of non-drainage in the whole test process except for the initial drainage and consolidation stage; the temperature path is 25–50–25 °C, and the pore water in the specimen is not discharged in the process of temperature increase/decrease and is discharged when the corresponding temperature is reached. Solidification. At the same time, the thermal consolidation characteristics of saturated clay under multi-stage heating/cooling cycle with temperature paths of 20–50–80–50–20 °C were also carried out.

4.1 Thermal Consolidation Response of Saturated Silty Clay Induced by Undrained Temperature Heating and Cooling Cycle

4.1.1 Overview of the Test

(1) Test sample

A kind of powdery clay was selected, air-dried, crushed, and sieved through 0.5 mm to prepare remodeled soil samples for the test. The basic physical properties of the soil were as follows: plastic limit ω_p was 17.6%, liquid limit ω_L was 31.4%, plasticity index $I_P = 13.8$, and relative density of the soil particles G_s was 2.71. Wet soil samples with a certain water content were prepared by mixing the soil with water and then compacted uniformly into a specimen with a height of 8 cm and a diameter of 3.91 cm in a saturated vessel in five layers. The dry weight ρ_d of the prepared

G. Yang and B. Bai, *Experimental and Theoretical Study on Thermal Consolidation Behavior of Saturated Clay*, https://doi.org/10.1007/978-981-96-6909-7_4

specimen was 15.5 kN/m^3. The specimen was saturated with distilled water under vacuum, the water content ω after saturation was 25.1%, and the degree of saturation S_r was 96.9%.

(2) Pilot program

When the specimen was installed, the temperature of the pressure chamber was heated to 25 °C (slightly greater than the ambient temperature of the test chamber, 23 °C) for isothermal consolidation test. When the consolidation was stabilized, the specimen was heated up under undrained conditions, and the power of the heating plate used was 200 W (when $\sigma_3 = 50$, 150 kPa, the heating rate was 0.26 °C/min, and when $\sigma_3 = 100$ kPa, the heating rate was 0.18 °C/min; the average value was 0.22 °C/min). In this way, the temperature of the saturated specimen was gradually increased from 25 to 75 °C (noted as 25–75 °C). Then, the temperature of the specimen was again reduced to 25 °C after (noted as 75–25 °C). This was repeated for a total of six temperature cycling action processes. As a comparison, a faster kind of temperature cycling action test with a faster rate of temperature increase (taking $\sigma_3 = 100$ kPa) was also carried out. The power of the heating plate used was 500 W (the heating rate was 2.01 °C/min).

4.1.2 Isothermal Consolidation Process After Temperature Increase

Figures 4.1, 4.2, and 4.3 give the drainage consolidation process of the saturated specimens under initial isothermal (i.e., 25 °C, heating plate power of 200 W) conditions when $\sigma_3 = 50$, 100, and 150 kPa. At the same time, the change of pore water pressure and volumetric strain with time is also given when the heating rate is large (i.e., $\sigma_3 = 100$ kPa, and the heating plate power is 500 W).

(1) Dissipation process of pore water pressure

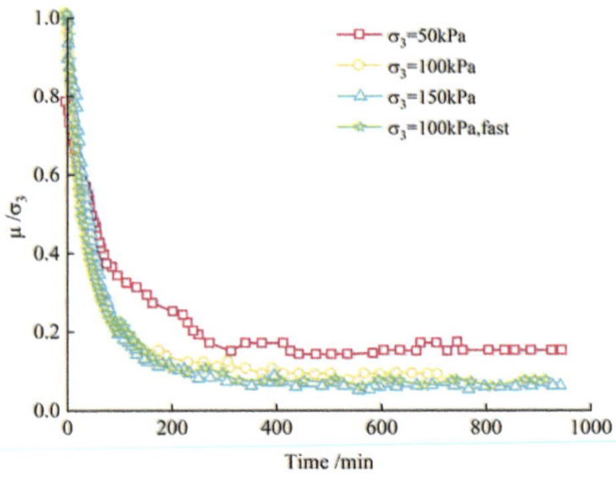

Fig. 4.1 Dissipation process of pore water pressure in saturated soil samples under initial temperature isothermal condition

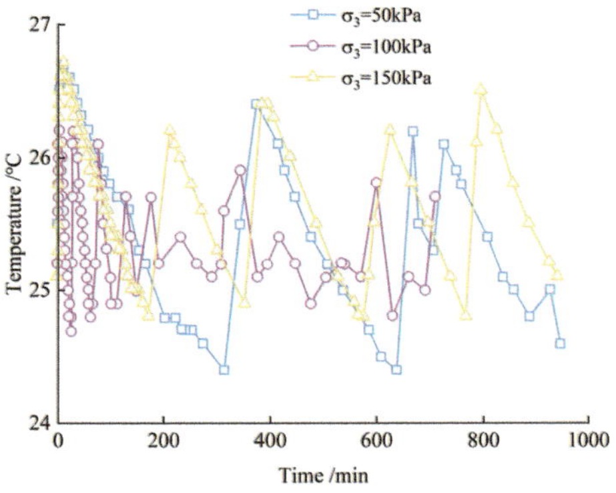

Fig. 4.2 Fluctuations in temperature of saturated soil samples under initial temperature isothermal conditions

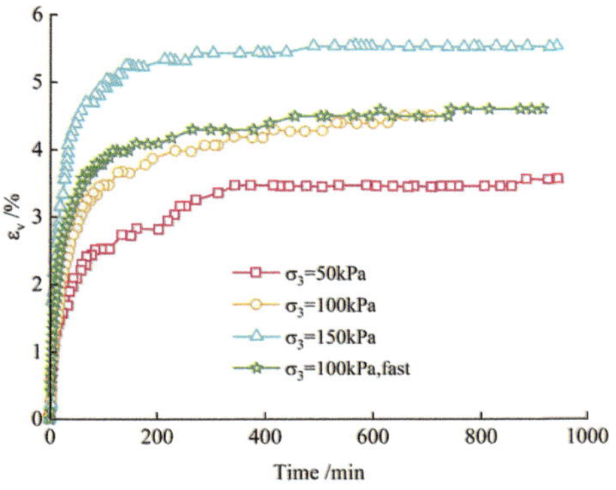

Fig. 4.3 Variation process of strain in saturated soil samples under initial temperature isothermal condition

As can be seen in Fig. 4.1, the trend of pore water pressure dissipation in the initial drainage consolidation process of the specimens under different peripheral pressures is consistent. In the initial stage of consolidation (e.g., within $t = 100$ min), the pore water pressure dissipates quickly, and the dissipation rate of pore water pressure increases with the increase of the peripheral pressure. When $t = 300$ min, the pore water pressure tends to stabilize, and then the pore water pressure in the specimen has some fluctuations due to the fluctuation of the specimen temperature. It can be seen that the dissipation process of pore water pressure under the two heating rate conditions is relatively consistent.

Figure 4.1 shows that the pore water pressure does not dissipate to zero during the isothermal consolidation process after the temperature increase (i.e., 25 °C), and a residual pore water pressure exists, which is related to the variability of the specimen temperature with respect to the ambient temperature. For example, when $t = 700$ min, the pore water pressures at $\sigma_3 = 50$, 100, and 150 kPa are 9, 11, and 10.5 kPa (i.e., $u/\sigma_3 = 0.18$, 0.11, and 0.07), respectively.

(2) Temperature fluctuation

As can be seen in Fig. 4.2, the temperature fluctuation is relatively large in the early stage of consolidation (e.g., $t = 500$ min), with the maximum value of 26.8 °C and the minimum value of 24.7 °C; while in the late stage of consolidation (e.g., $t = 500$ min), the frequency and amplitude of temperature fluctuation are obviously reduced. The fluctuation of the sample temperature is related to the continuous adjustment of the temperature of the water in the pressure chamber due to the continuous discharge of the pore water in the sample, especially in the early stage, the temperature fluctuation is more frequent because more water with a slightly lower temperature enters into the pressure chamber for compensation.

(3) Changes of volume strain

As can be seen in Fig. 4.3, after the specimen consolidation and stabilization (e.g., $t = 700$ min), the specimen volume strains are $\varepsilon_v = 3.54\%$, 4.48%, and 5.52%, respectively, when $\sigma_3 = 50$, 100, and 150 kPa. Obviously, the volumetric strain of the specimen increases with the increase of the peripheral pressure.

4.1.3 Temperature Changes Under Undrained Conditions

(1) Analysis of the temperature evolution process

Figure 4.4 shows the temperature change of the specimen under the action of two kinds of pressure (i.e., $\sigma_3 = 50$, 150 kPa; heating rate of 0.26 °C /min) without drainage conditions, the temperature action of the specimen in the temperature of the temperature of the process of warming in each cycle (25–75 °C, for example), which is the first stage, the duration of 180 min. It can be seen that, in the beginning of the 100 min of warming, the rate of increase in temperature is relatively fast. It can be seen that the rate of temperature increase is relatively fast during the first 100 min of warming, and then the rate of warming slows down. Obviously, the trend of temperature change under different pressurization is almost the same, and the trend of temperature change of the specimen is also the same in each cycle of warming process.

In the second stage, the specimens were allowed to remain undrained under isothermal conditions (i.e., 75 °C) for 450 min. As can be seen in Fig. 4.4, the fluctuation range of the temperatures (i.e., 74.4–75.6 °C) was relatively small under different enclosure pressures and number of cycles, and the fluctuation tendency was also almost the same.

In the third stage, the specimen temperature was reduced from 75 to 25 °C (i.e., 75–25 °C) for 240 min. It can be seen that in the initial stage of temperature reduction (e.g., within $t = 50$ min), the temperature of the specimen decreased relatively quickly, and then the rate of temperature reduction of the specimen decreased significantly due to the reduction of the difference between the temperature in the pressure chamber and the ambient temperature. It can also be seen that the temperature change trend of the specimen is almost the same for different pressures and cycle times.

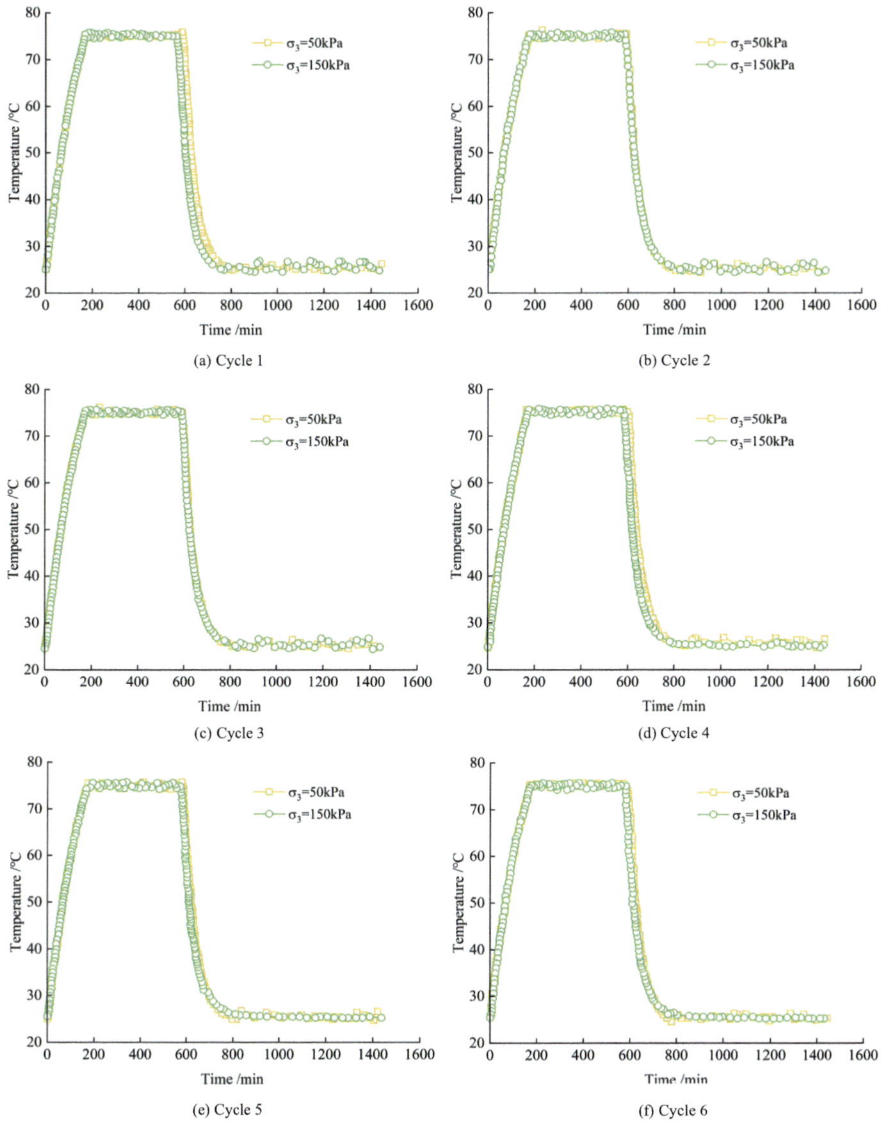

Fig. 4.4 Variations of temperature under repeated heating–cooling processes

In the fourth stage, the specimens were kept undrained under isothermal conditions (i.e., 25 °C) for 600 min. It can be seen that the range of temperature fluctuations (i.e., 24.5–25.5 °C) was small for different enclosure pressures and number of cycles. Overall, the change rule of temperature is almost the same throughout the change process of temperature loading cycle (Fig. 4.4, the time required is 1470 min) and under different enclosure pressures as well as different number of cycles.

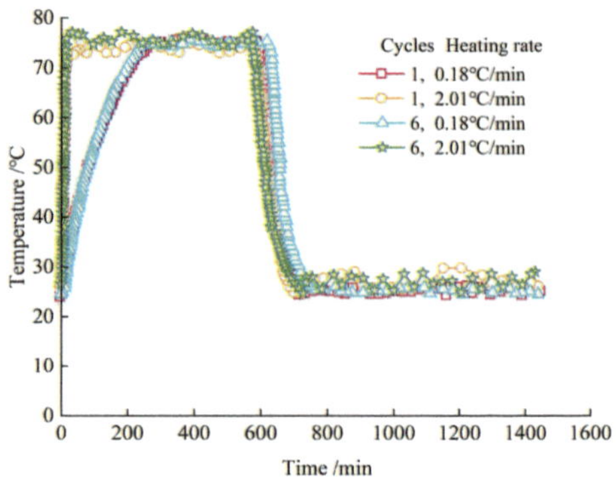

Fig. 4.5 Variations of temperature under various heating ratios

(2) Effect of warming rate on the temperature evolution process

Figure 4.5 shows the changes of the specimen temperature during the first and sixth temperature loads at two heating rates (i.e., 0.18 °C/min and 2.01 °C/min; $\sigma_3 =$ 100 kPa). It can be seen that the time required to increase the specimen temperature from 25 to 75 °C is 277.8 min when the heating rate is 0.18 °C/min, while the time required is only 24.9 min when the heating rate is 2.01 °C/min.

4.1.4 *Pore Water Pressure Changes Under Undrained Conditions*

(1) Effect of the number of temperature cycles on pore water pressure

Figure 4.6 gives the course of normalized pore water pressure with time under cyclic temperature loading when the enclosing pressure $\sigma_3 = 50$, 100, and 150 kPa, respectively (i.e., 25 °C–75 °C–25 °C). It can be seen that the temperature-induced pore water pressure increases continuously with increasing temperature (i.e., 25–75 °C) in the first stage, and the growth trend of pore water pressure within each cycle is similar. At the same perimeter pressure (e.g., $\sigma_3 = 50$ kPa), the normalized maximum pore water pressure increases slightly with the number of cycles and gradually becomes equal after the third temperature cycle.

During the isothermal (i.e., 75 °C) undrained process, the fluctuation of specimen temperature caused some fluctuation of the pore water pressure, but the pore water pressure basically remained at a constant value. During the cooling process (75–25 °C), the pore water pressure within the specimen also decreased rapidly and gradually stabilized, but did not drop to zero; i.e., there was a significant residual pore water pressure. However, as the number of temperature cycles increases (e.g., $N = 1, 2, 3, 4, 5, 6$), the rate of decrease in normalized pore water pressure appears to slow down, and the residual pore water pressure increases when the temperature is reduced to the set temperature (i.e., 25 °C). As can be seen from the comparison

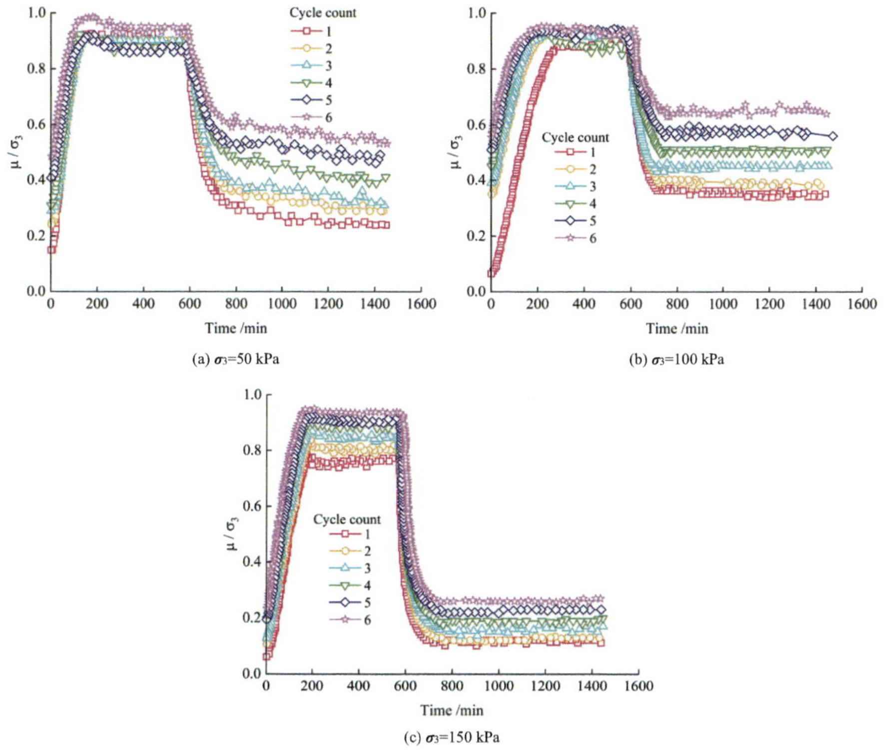

Fig. 4.6 Effects of cyclic number of temperature on pore pressure

of Fig. 4.6a–c, this trend diminishes with increasing perimeter pressure (i.e., $\sigma_3 =$ 50, 100, 150 kPa), and the normalized residual pore water pressure also tends to decrease.

(2) Influence of ambient pressure on pore water pressure

Figure 4.7 shows the variation of normalized pore water pressure with time during the first and sixth temperature cycles of the specimens when the ambient pressure σ_3 = 50 kPa and 150 kPa. From Fig. 4.7, it can be seen that the normalized maximum pore water pressure tends to decrease with the increase of the surrounding pressure; however, the actual pore water pressure increases significantly. For example, during the first temperature cycle, when $\sigma_3 = 50$ kPa and 150 kPa, the normalized pore water pressures were $u/\sigma_3 = 0.95$ and 0.80 (i.e., $u = 47.5$ kPa and 120.0 kPa), respectively, after the temperature was increased to the set temperature (i.e., 75 °C). Moreover, the normalized residual pore water pressure tends to decrease with the increase of the surrounding pressure after the temperature is reduced to the set value (i.e., 25 °C), while the actual pore water pressure increases. For example, the normalized residual pore water pressure is $u/\sigma_3 = 0.24$ and 0.11 (i.e., $u = 12.0$ kPa and 16.5 kPa) at σ_3 = 50 kPa and 150 kPa, respectively, during the first temperature cycling.

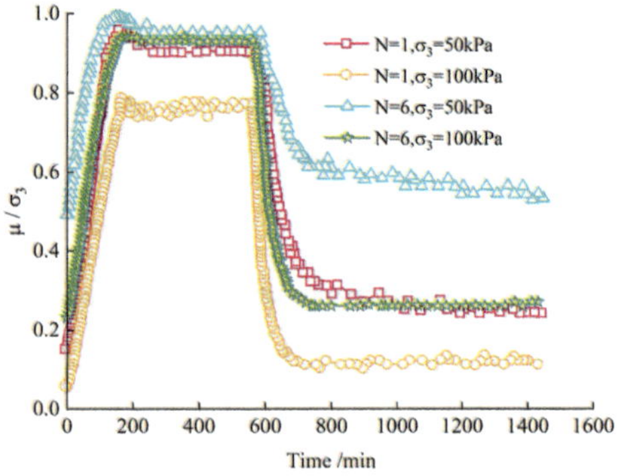

Fig. 4.7 Variations of pore pressure under various confining pressures

(3) Effect of warming rate on pore water pressure

Figure 4.8 shows the pore water pressure changes with time at two heating rates (i.e., 0.18 °C/min and 2.01 °C/min, respectively) during the first and sixth temperature cycles when the peripheral pressure $\sigma_3 = 100$ kPa. It can be seen that the faster the warming rate, the faster the increase rate of pore water pressure. Figure 4.8 shows that the normalized pore water pressures at the completion of warming are very close to each other, despite the different warming rates, when the specimen temperature is increased to 75 °C. The normalized pore water pressures at the completion of warming are very close to each other. For example, for the first temperature cycling process, the maximum value is $u/\sigma_3 = 0.91$. In fact, the growth trend of the pore water pressure in the specimen is consistent with the temperature trend (Refer to Fig. 4.5).

However, when the pore water pressure reaches the peak value, the pore water pressure inside the specimen basically remains near the maximum value of pore water pressure when the heating rate is relatively slow (i.e., 0.18 °C/min); and when the heating rate is relatively fast (i.e., 2.01 °C/min), the pore water pressure inside the specimen reaches the maximum value followed by an obvious decreasing tendency,

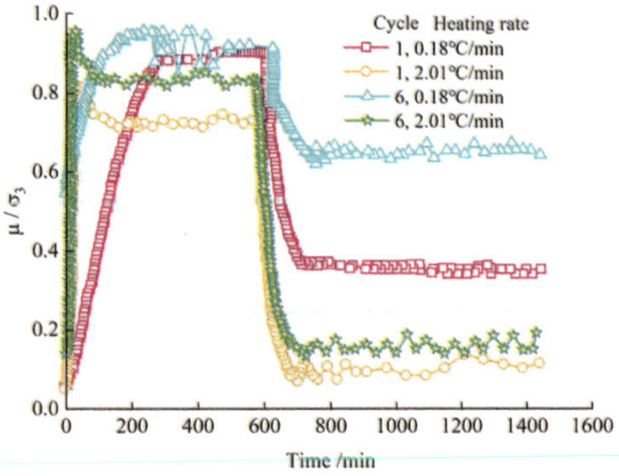

Fig. 4.8 Variations of pore pressure under various heating ratios

which is only maintained at a stable value (i.e., $u/\sigma_3 = 0.73$). In fact, the expansion rates of solid particles and pore water within the specimen are not completely synchronized under different heating rates, and the heating of the specimen sidewalls (or ends) and the interior of the specimen is not uniform. Therefore, there is also a sample from the outside to the inside of the heat exchange process and pore water flow process, and gradually converge to the equilibrium state, so that the sample of different locations and different moments of the pore water pressure changes in the process is very complex, resulting in the sample to reach a stable value of pore water pressure (measured from the bottom of the sample) appears to be a big difference.

The residual pore water pressure also varied significantly when the temperature of the specimen decreased from 75 to 25 °C. The temperature increase rate was higher during the warming phase than during the warming phase. The residual pore water pressure of the specimen with faster heating rate (i.e., $u/\sigma_3 = 0.11$) is much smaller than that of the specimen with slower heating rate (i.e., $u/\sigma_3 = 0.36$) during the warming stage. Moreover, this difference seems to be more pronounced as the number of temperature cycles increases (e.g., when the number of cycles $N = 6$, refer to Fig. 4.8).

4.1.5 Volume Strain Changes

During undrained heating or cooling, the pore water within the specimen is not discharged. However, due to the thermal expansion or contraction effects of the solid particles and pore water in the specimen, the specimen also experiences a significant volume change. Define the volumetric strain in the specimen due to temperature change as $\varepsilon_v = \Delta V/V_0$ (%, V_0 is the initial volume of the specimen and ΔV is the volumetric change in the specimen due to temperature change). This volume change can be indirectly determined by the volume change of water in the pressure chamber. That is, according to the thermal expansion coefficient of water ($\alpha_w = 2.0 \times 10^{-4}$/°C) will be the pressure chamber of the volume change of water (test measured at the initial temperature, that is, under the condition of 25 °C) into the pressure chamber at a certain temperature (heating or cooling process, such as 75 °C) the volume change of water, you can know that the temperature change in the volume of the specimen. Figure 4.9 gives when $\sigma_3 = 50$, 100, 150 kPa (the first temperature cycle process) of the specimen volume strain with time. When $\sigma_3 = 50$ kPa and 150 kPa specimen heating rate is slightly faster (0.26 °C/min), and when $\sigma_3 = 100$ kPa specimen heating rate is slightly slower (0.18 °C/min).

As can be seen from Fig. 4.9, during the temperature increase of the specimen (i.e., 25–75 °C), the solid particles and pore water inside the specimen expand in volume, thus causing a significant thermal volumetric strain in the specimen. When the temperature is increased to the set temperature (75 °C), the thermal volumetric strain of the specimen also increases to the maximum value, after which the thermal volumetric strain of the specimen fluctuates somewhat with the fluctuation of temperature. For example, when $\sigma_3 = 50$ kPa, the volumetric strain is $\varepsilon_v = 3.6\%$ (positive

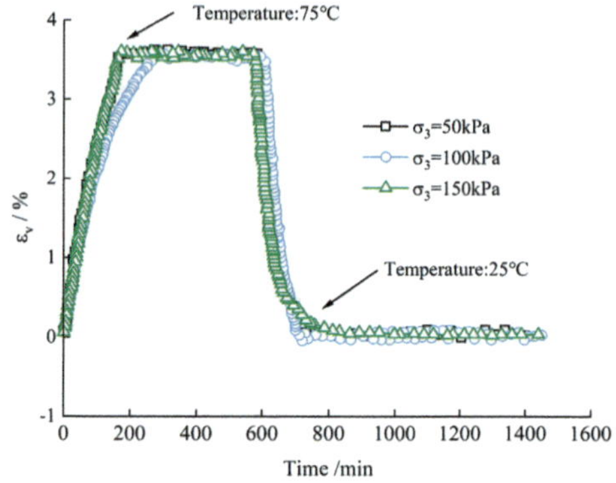

Fig. 4.9 Volumetric strains induced by heating–cooling cycles

value indicates expansion deformation). When the specimen temperature drops from 75 to 25 °C, the volumetric strain of the specimen decreases again due to the volume contraction of the solid particles and pore water inside the specimen and finally approaches zero, which is a reversible volumetric change process. It can be seen that the thermal volumetric strain change trend of the specimen is consistent with the temperature change process. Figure 4.9 shows that the size of the circumferential pressure has almost no effect on the thermal volumetric strain of the specimen; at this time, the thermal expansion effect dominates.

4.1.6 Summary

(1) During the whole change process of repeated warming-cooling of saturated pulverized clay, the temperature change process is almost the same for different enclosing pressures and different numbers of temperature cycles.
(2) The temperature-induced pore water pressure increases continuously with increasing temperature, and the growth trend of pore water pressure during each temperature cycle is similar. The normalized pore water pressure increased slightly with the number of temperature cycles under the same enclosure pressure, but gradually equalized after the third temperature cycle.
(3) During the cooling process, the pore water pressure in the specimen decreases rapidly and stabilizes, but there is a significant residual pore water pressure. As the number of temperature cycles increases, the rate of decrease in normalized pore water pressure seems to slow down and the residual pore water pressure increases. In addition, there is a tendency for the normalized residual pore water pressure to decrease as the enclosing pressure increases.
(4) When the specimen temperature decreases, the residual pore water pressure of the specimen with a faster rate of warming during the warming phase is much smaller than that of the specimen with a slower rate of warming. This difference appears to be more pronounced as the number of temperature cycles increases.

(5) Volume expansion of solid particles and pore water within the specimen occurs as a result of the temperature increase. Afterward, as the temperature decreases, the volume of solid particles and pore water within the specimen shrinks again, i.e., the volumetric strain of the specimen decreases and tends to zero, which is manifested as a reversible volumetric change process.

4.2 Thermal Consolidation Characteristics of Saturated Silty Clay Caused by Single-Stage Heating and Cooling Cycles

4.2.1 Overview of the Test

(1) Test Soil Sample

The soil selected for the experimental study was a kind of powdery clay commonly found in the northern region. The soil was air-dried, crushed, and sieved through 0.5 mm sieve to prepare remodeled soil samples for testing. The basic physical properties of the soil were measured in the indoor test: the plastic limit ω_p was 17.6%, the liquid limit ω_L was 31.4%, the plasticity index I_p was 13.8, the specific gravity of the soil particles G_s was 2.66, and the dry weight ρ_d was 15.5 kN/m^3.

The soil samples were prepared into wet soil samples by hitting the sample method with a good moisture content and sufficiently wet and uniform soil and poured into the compactor equipped with a three-valve membrane in five layers, and each layer was compacted 15 times and evenly compacted to a height of 8 cm and a diameter of 3.91 cm, resulting in a soil sample with a density of 1.96 g/cm^3, and the moisture content was measured to be 23.7%, which is a relatively small difference from the predetermined 24% moisture content in the design of the experiment. After all soil samples were prepared, the specimens were saturated with distilled water by vacuum pumping, and the water content after saturation was 25%, and the saturation degree S was 98%.

(2) Test program

The self-developed thermal consolidation test device was utilized to carry out six consecutive cyclic tests with temperatures ranging from 25 °C to 50 °C, in which the specimen peripheral pressures were 50 kPa, 100 kPa, and 150 kPa, respectively, and the specific test scheme is shown in Table 4.1. Through the test phenomena, the evolution of temperature, pore water pressure, and strain of the solidified body during thermal consolidation under different peripheral pressures was analyzed by the number of cycles, and the study of thermal property mechanism was applied to explain the evolution of pore water pressure and strain of the solidified body.

Table 4.1 Single-stage cyclic temperature rise/fall test program

Programmatic	Pressurization (kPa)	Initial temperature (°C)	Increase the temperature (°C)	Lower the temperature (°C)	Number of cycles
1	50	25	50	25	6
2	100	25	50	25	6
3	150	25	50	25	6
Clarification	No drainage during warming/cooling, and drainage for consolidation after warming to 50 °C and cooling to 25 °C				

4.2.2 *Effect of the Number of Heating–Cooling Cycles on Pore Water Pressure*

Figure 4.10 gives the process of normalized pore water pressure with time under cyclic temperature loading when the enclosing pressure $\sigma_3 = 50$, 100, and 150 kPa, respectively (25 °C–50 °C–25 °C, for example). It can be seen that the pore water pressure due to temperature rise decays to a certain extent with the increase in the number of cycles, but this decay process mainly occurs in the first 3 cycles. Starting from the fourth cycle, the substituent pore water pressure generated by increasing the temperature to 50 °C is almost the same as that of the third cycle, and no further attenuation occurs. Therefore, the study was mainly carried out for the first 3 cycles.

Taking the peak value of the pore water pressure that can be reached at the end of the warming process as the object of study, the pore water pressure/pressure produced by the specimen with an enclosing pressure of 50 kPa in the first cycle was 0.68, the ratio in the second cycle was 0.48, and the pore water pressure/enclosure pressure produced in the third cycle was 0.3. From the fourth to the sixth cycles, the ratios were 0.36, 0.38, and 0.38, respectively.

In a further analysis of the pore water pressure, it can be seen that the pore water pressure reaches its peak when the temperature increases to 50 °C. The statistics of the $(u/\sigma_3)_{\text{peak}}$ data at different periapsis pressures are shown in Table 4.2.

It can be clearly seen from this set of data that from the first cycle to the second, the ratio undergoes a 29% decay, from the second cycle to the third, a 20% decay occurs, and from the fourth cycle onwards, almost no decay occurs and the value fluctuates within a small range. In other words, as the number of cycles increases, the rate of decay decreases until no decay occurs. Meanwhile, the data in Table 4.2 also further illustrates that for the same number of cycles, the larger the enclosing pressure is, the smaller the $(u/\sigma_3)_{\text{peak}}$ is, which is consistent with the analysis above.

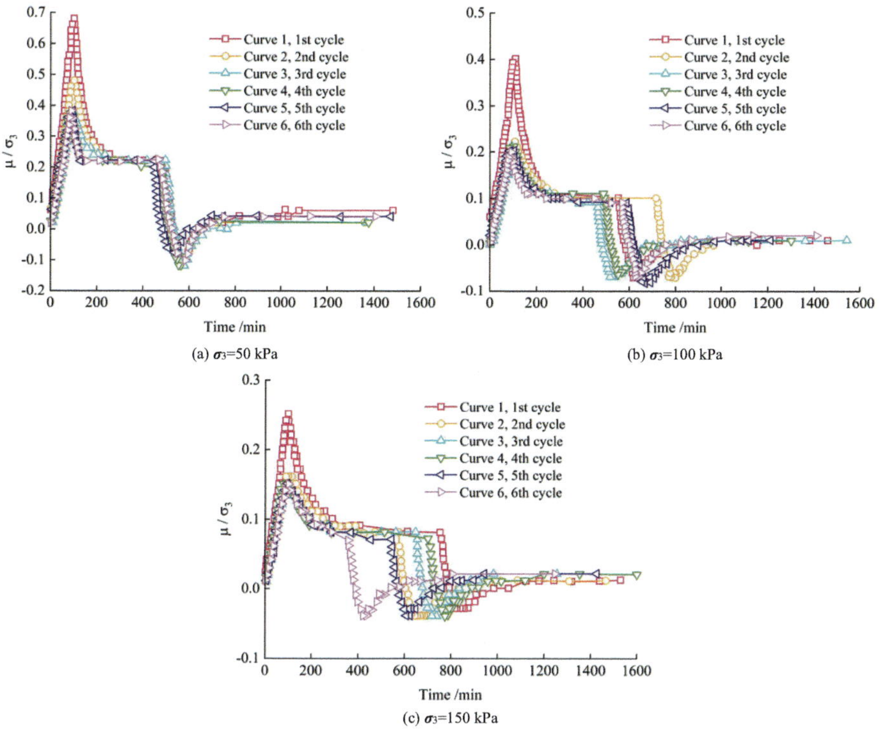

Fig. 4.10 Evolution of pore water pressure under the effect of temperature in a single-stage ramp-up/down cycle

Table 4.2 A $(u/\sigma 3)_{\text{peak}}$ attained by warming up to 50 °C with different number of cycles at different enclosure pressures

Perimeter pressure (kPa)	Number of cycles					
	1	2	3	4	5	6
50	0.68	0.48	0.38	0.36	0.38	0.38
100	0.40	0.22	0.20	0.20	0.21	0.20
150	0.25	0.16	0.15	0.15	0.14	0.15

4.2.3 Effect of the Number of Heating–Cooling Cycles on Volume Strain

Figure 4.11 gives the changes in the solidified body strain induced by cyclic temperature increase/decrease in the temperature path 25–50–25 °C under three different peripheral pressures (50, 100, 150 kPa). Comparative analysis shows that the body strain occurring in the specimen consolidated at 50 °C at high temperature is significantly larger than that occurring in the consolidated at room-temperature (25 °C) condition. High-temperature consolidation is the process in which the pore water in the specimen is discharged from the specimen to shrink the volume of the specimen, while room-temperature consolidation is the process in which the pore water in the specimen absorbs the water and the volume of the specimen expands. In the same cycle of the test, volume contraction is greater than the volume of expansion, so a cycle after the completion of the specimen volume shows a certain amount of contraction.

The solidus strain measured by the change in water volume inside the specimen shows a certain degree of attenuation of the deformation as the number of cycles increases. This decay process is most evident during the first 3 cycles. Starting from the fourth cycle, the variation curves of the solidus strain produced during high-temperature (50 °C) consolidation are almost the same as that of the third cycle, and no further attenuation phenomenon occurs. Taking the solidus strain measured by the change of water volume inside the specimen during the high-temperature (50 °C) drainage consolidation after the warming process as the object of study, the specimen with an enclosing pressure of 50 kPa produced a solidus strain of 2.60% in the first cycle, the deformation of this deformation was 2.40% in the second cycle,

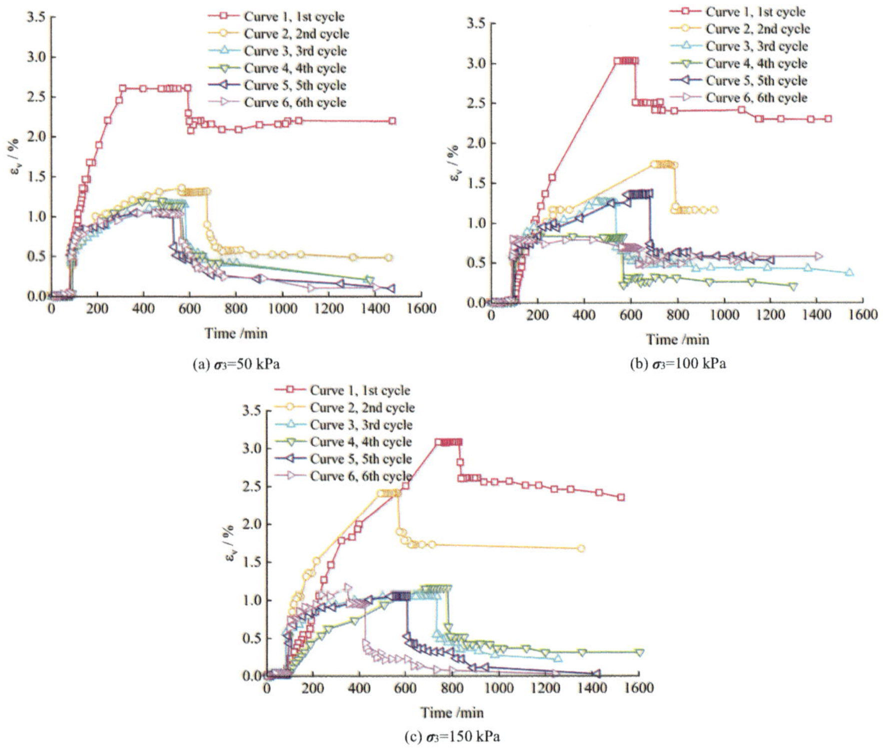

Fig. 4.11 Evolution of solidus strain under the temperature action of a single-stage ramp-up/down cycle

Table 4.3 Body strains achieved by warming to 50 °C with different number of cycles at different circumferential pressures

Perimeter pressure (kPa)	Number of cycles					
	1	2	3	4	5	6
50	2.60	1.30	1.15	1.15	1.04	1.04
100	3.02	1.72	1.25	0.83	1.25	0.78
150	3.07	2.40	1.04	1.15	1.04	0.94

and 1.15% in the third cycle, and the deformation of this deformation was 1.15%, 1.04%, 1.04%, and 1.04%, respectively, from the fourth to the sixth cycle. 1.15%, 1.04%, and 1.04%.

The statistics of the solidus strain data at 50 °C after warming under different peripheral pressures can be referred to Table 4.3. It can be clearly seen from this set of data that, for example, with the data of peripheral pressure of 50 kPa, from the first cycle to the second cycle, the value undergoes 50% decay, and from the second cycle to the third cycle, it undergoes 11.5% decay, and the rate of decay decreases from the fourth cycle onward, and almost no decay, and the values fluctuated within a small range. In other words, as the number of cycles increases, the rate of decay decreases until no decay occurs. Meanwhile, the body strain of the specimen with an enclosing pressure of 100 kPa is larger in the fifth cycle, which is attributed to the fact that the other high-temperature consolidation usually takes about 400 min, while this consolidation process lasts for 496 min. In addition, the precision of the test equipment is also an important factor that affects the test results.

4.3 Thermal Consolidation Characteristics of Saturated Silty Clay Caused by Multi-stage Heating and Cooling Cycles

4.3.1 Overview of the Test

(1) Test soil sample

A moderately plastic saturated clayey soil was selected for the test. Disturbed soil samples were prepared by air-drying, crushing, and sieving through 0.5 mm sieve. The average values of the basic physical properties of the test soil samples were: plastic limit ω_p was 15.8%, liquid limit ω_L was 34.9%, plasticity index I_p was 19.1, and the specific gravity of the soil particles G_s was 2.71. The specimen size φ was 3.91 cm in diameter, 8 cm in height h, and the prepared water content w was 29.7%, and the dry weight degree ρ_d was 15.1 kN/m^3. The specimens were saturated by vacuum pumping with distilled water to a saturation of $S_r > 95.1\%$.

(2) Test program

The main study is to investigate the thermal consolidation response of saturated soil samples during repeated changes in hot–cold temperature. The specimen first completes the main consolidation under a certain circumferential pressure σ_3 (50, 100, 150 kPa, refer to Table 4.4), and then the temperature of the specimen is increased by the change of the water temperature in the pressure chamber under the undrained condition (i.e., it is increased from 20 to 50 °C, recorded as 20–50 °C), and at the same time, the body strains of the specimen are determined under the undrained condition due to the thermal expansion effect of the specimen's solid particles and the pore water. Caused by the thermal expansion effect of the solid particles and pore water of the specimen, the body strain of the specimen. At this point, the specimen was allowed to consolidate under the drained condition, the pore water pressure was dissipated, and the specimen was thermally consolidated and deformed. Subsequently, the drain valve was closed and the temperature of the specimen was continued to be increased (i.e., from 50 to 80 °C, noted as 50–80 °C), while the thermal consolidation response of the specimen was determined during the further increase in temperature.

When the specimen temperature reaches 80 °C and complete solidification, graded reduction in the temperature of the specimen (i.e., from 80 to 50 °C, recorded as 80–50 °C and from 50 to 20 °C, recorded as 50–20 °C). In the temperature reduction process, the specimen is not drained state, the temperature down to the specified value and then open the drain valve to let the specimen drainage (or water absorption), and then the next stage of the temperature reduction process. In this way, the specimen to complete a temperature cycle and solidification process.

Then, the second temperature cycle change process, this time the temperature graded increase and graded reduction of the test process and the first temperature cycle process is the same, a total of two temperature cycle process.

4.3.2 *Pore Pressure and Thermal Consolidation Deformation Caused by Temperature Rise in Multi-stage Circulation*

(1) Heating process

As shown in Fig. 4.12, when the peripheral pressure $\sigma_3 = 50$ kPa, the first and second temperature cycle processes of the two temperature rise stage of the specimen temperature change process. It can be seen that the rise in specimen temperature from 20 to 50 °C (i.e., the upward amplitude of 30 °C) is faster than the subsequent rise in temperature from 50 to 80 °C (also with an upward amplitude of 30 °C), with the former taking about 94 min and the latter taking about 240 min (refer to Table 4.4). Overall, the rate of temperature increase gradually slowed down as the temperature increased (i.e., throughout the entire continuum of temperature increase from 20 to 50 °C and later to 80 °C). It should be noted that the second cycling process was almost identical to the first cycling process, even though the specimens had already gone through the complex process of temperature increase, decrease, and drainage consolidation during the first temperature cycling process.

Table 4.4 Pore pressure induced by heating in undrained conditions

Pressurization σ_3/kPa	First cycle 20–50 °C		First cycle 50–80 °C		Second cycle 0–50 °C		Second cycle 50–80 °C	
	Stabilization time/min	Pore water pressure/kPa	Stabilization time/min	Pore water pressure/kPa	Stabilization time/min	Pore water pressure/kPa	Stabilization time/min	Pore water pressure/kPa
50	94	36	240	46	94	31	240	46
100	99	54	245	70	99	29	245	50
150	96	51	249	82	96	47	249	71

Fig. 4.12 Temperature variations of the specimens due to heating or cooling in undrained conditions

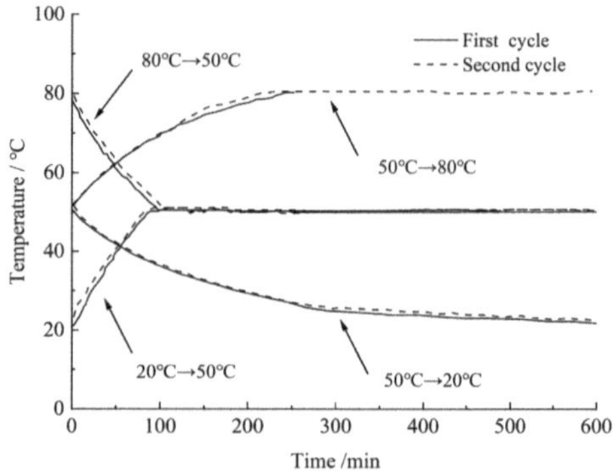

The tests show that when σ_3 is 100 kPa and σ_3 is 150 kPa, the variation of temperature rise during the two temperature cycles obtained is almost the same as that when σ_3 is 50 kPa, i.e., the heating process of the specimen is independent of the surrounding consolidation pressure on the specimen, which can be seen from the time required for the heating process in several cases given in Table 4.4.

(2) Heating-induced pore water pressure

As shown in Fig. 4.13, two temperature cycling processes as well as two heating stages (i.e., 20–50°C and 50–80 °C) of the pore water pressure of the specimen under the conditions of three different enclosure pressures (50, 100, and 150 kPa) were carried out.

As can be seen in Fig. 4.13, significant pore water pressure is generated within the specimen as the temperature increases. Furthermore, for the same cycling process and at the same stage, as the enclosing pressure increases (e.g., Fig. 4.13c compared to Fig. 4.13a), the pore water pressure increases accordingly. For the same pressure and the same cycling process, the pore water pressure generated by increasing the temperature from 20 to 50 °C is slightly smaller than that generated by increasing the temperature from 50 to 80 °C. However, the temperature rise process in the former stage (i.e., before reaching the set temperature value) is almost linear and slightly faster, which is consistent with the faster temperature rise process in the corresponding stage given in Fig. 4.12.

In addition, the temperature-induced pore water pressures at both temperature rise stages of the second temperature cycle were smaller than those at the corresponding stages of the first cycle. This phenomenon actually reflects an aging effect (or attenuation effect) of the saturated clayey soil during the cyclic process of temperature increase and decrease. In particular, the difference in pore water pressures generated by different cyclic processes is more obvious as the circumferential pressure increases. For example, when σ_3 is 150 kPa, the difference in pore water pressure between the two stages stabilized in the first cycle is $\Delta u = 31$ kPa, and the difference in pore water pressure between the two stages stabilized in the second cycle is $\Delta u = 24$ kPa (obtained from Table 4.4); however, when $\sigma_3 = 50$ kPa, the difference in

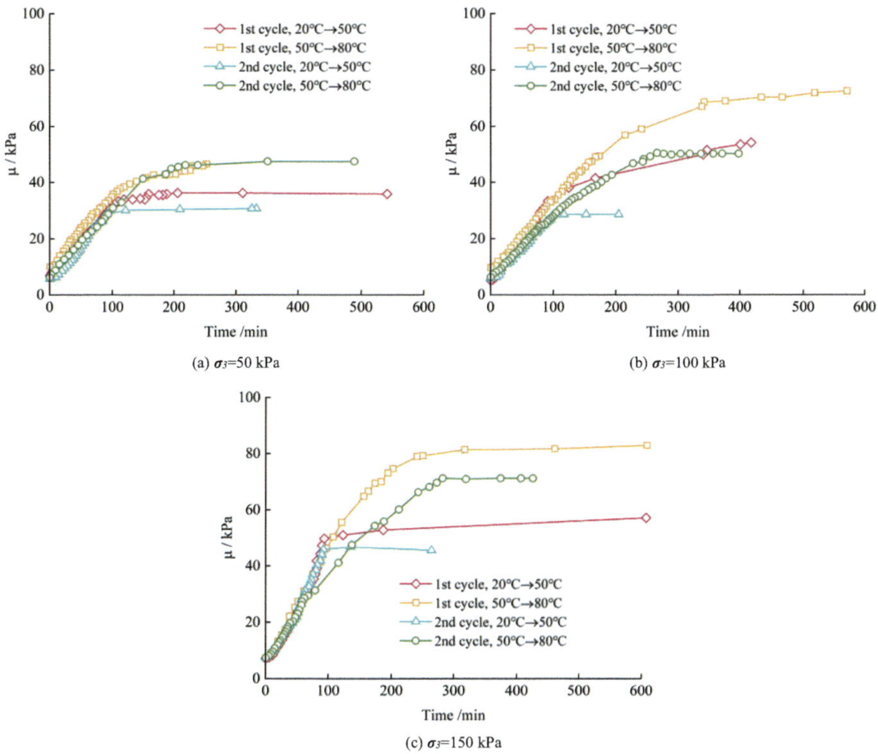

Fig. 4.13 Pore pressure induced by heating in undrained conditions

the pore water pressures induced in the two stages is smaller (10 kPa and 15 kPa in the two cyclic processes, respectively).

(3) Thermal consolidation deformation under drained conditions after warming up

Figure 4.14 gives the drainage consolidation process under isothermal conditions after temperature increase in two heating stages (i.e., 20–50 °C and 50–80 °C) in two temperature cycling processes under three different circumferential pressures (50 kPa, 100 kPa, and 150 kPa).

It can be seen that, due to the dissipation of the pore water pressure, under the same cycle, the strain of the consolidated drainage body under isothermal conditions after the specimen temperature is increased from 50 to 80 °C (the positive value indicates the volumetric shrinkage) is greater than that of the consolidated drainage body under isothermal conditions after the specimen temperature is increased from 20 to 50 °C, and the strain of the consolidated drainage body in the process of the second cycle should be correspondingly less than that of the consolidated drainage body in the first cycle. The strain of the consolidated drainage body during the second cycle should be smaller than that during the first cycle.

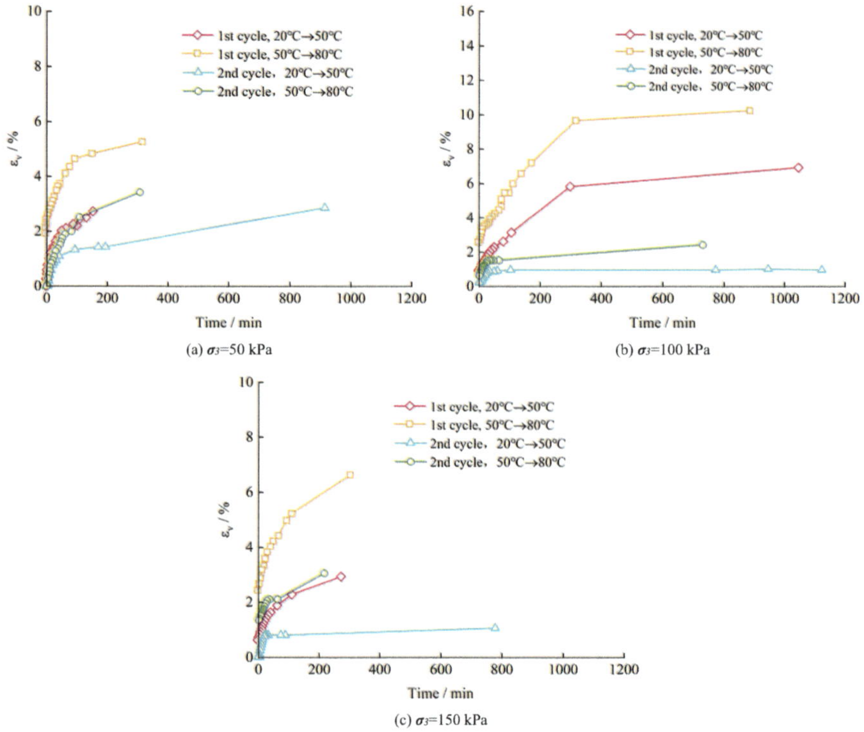

Fig. 4.14 Volumetric strain under isothermal and drained conditions after temperature rising

4.3.3 Thermal Consolidation Characteristics During Temperature Reduction in Multi-stage Cycles

(1) Temperature reduction process

As can be seen from Fig. 4.12, the temperature drop from 80 to 50 °C (downward drop value of 30 °C), the rate of temperature drop is faster than the rate of temperature drop from 50 to 20 °C (downward drop value is also 30 °C). This process is the opposite of the speed of the process of temperature increase in the corresponding stage.

The tests showed that, similar to the temperature rise process given in Fig. 4.12, the temperature fall process of the specimen was also independent of the enclosing pressure (50, 100, 150 kPa) and the number of temperature cycles (e.g., first, second).

(2) Pore water pressure during temperature reduction

In the temperature cycling process, the thermal contraction effect of solid particles and pore water within the specimen caused by the temperature drop under undrained conditions will inevitably induce changes in the pore water pressure and volume changes within the specimen.

As can be seen from Fig. 4.15, in the temperature drop process, because the cold contraction of the solid particles is smaller than the cold contraction of the pore water, so the specimen first produces a negative pore water pressure. When the temperature drops to roughly the set temperature value (e.g., when σ_3 is 50 kPa, the temperature drops from 80 to 50 °C in the first cycle, and the temperature drop time is 100 min, refer to Fig. 4.15), the pore water pressure starts to rise again, and there is a clear turnaround, which eventually produces a larger positive pore water pressure. Analyzing the reason for this, it can be concluded that when the temperature reaches a constant value, due to the coupling of temperature load and perimeter pressure load, the solid particles and pore water will be continuously adjusted, so that the pore water pressure generated by the perimeter pressure will eventually dominate. From this point of view, the thermal consolidation effect of saturated soil caused by the cooling process is irreversible.

With the continuation of time, the pore water pressure gradually stabilized (Fig. 4.15). Overall, it seems that the process of pore water pressure change is slightly more complicated under different number of temperature cycles, different cooling stages, and different conditions of the surrounding pressure during the cooling process. In general, with the increase of the surrounding pressure, the negative pore

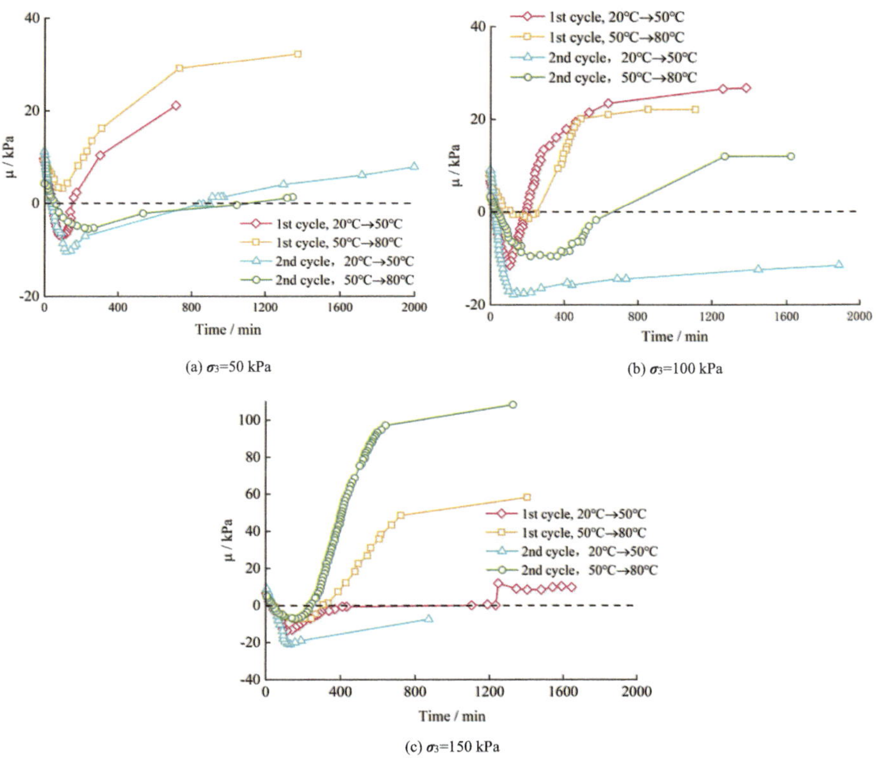

Fig. 4.15 Pore pressure induced by cooling in undrained conditions

water pressure caused by the cooling process in the same cycle stage and the same cooling stage increases, and the positive pore water pressure generated by the conversion of the negative and positive pore water pressure is also greater. Comparison of Fig. 4.15 with Fig. 4.13 shows that the time required for the pore water pressure to reach a stable value during the cooling process is much longer than the time required for the pore water pressure to reach a stable value during the warming process.

(3) Thermal consolidation deformation under post-cooling drainage conditions

After cooling, the same consolidation drainage (or water absorption) process was carried out (Fig. 4.16). As can be seen from Fig. 4.16, when the enclosing pressure is small (e.g., Fig. 4.16a, $\sigma_3 = 50$ kPa), the specimens all show the drainage effect; when the enclosing pressure is large (Fig. 4.16 b and c), $\sigma_3 = 100$ kPa and $\sigma_3 = 150$ kPa), the majority of the stages still show the drainage effect, whereas the individual stages show the water absorption effect (the second temperature cycle process when the temperature decreased from 80 to 50 °C). This is related to the complex coupling of external loads (enclosure pressure) and temperature loads (warming or cooling), a phenomenon that also tends to become more complex with the number of temperature cycles. A large number of tests are needed to verify this point. This phenomenon actually reflects the irreversibility of the thermal consolidation process in saturated soils, and is a manifestation of the temperature-induced "aging effect" of saturated soils.

Overall, for each stage of temperature change, the amount of water drained after cooling was much smaller compared to the amount of water drained from the heated specimen (Fig. 4.14) (comparison of Fig. 4.14a and Fig. 4.16a).

4.3.4 Summary

The thermal consolidation characteristics of saturated clayey soils during multi-stage temperature increase–decrease cycles were tested by a self-developed temperature-controlled thermal consolidation test setup. Some conclusions were obtained as follows:

(1) The undrained heating process induces significant pore water pressure due to the difference in the thermal expansion effects of the solid particles and pore water of the specimen. The temperature-induced pore water pressure during the second cycle is less than that induced during the first cycle. This phenomenon actually reflects an aging effect of the saturated clay soil during the cyclic process of increasing and decreasing temperature.
(2) During the decrease of temperature, negative pore water pressure is first generated within the specimen. When the temperature drops to the set temperature value, the pore water pressure starts to rise again with a significant turnaround, and then a positive pore water pressure will be generated. This is caused by the continuous adjustment of the thermal expansion effect of solid particles and

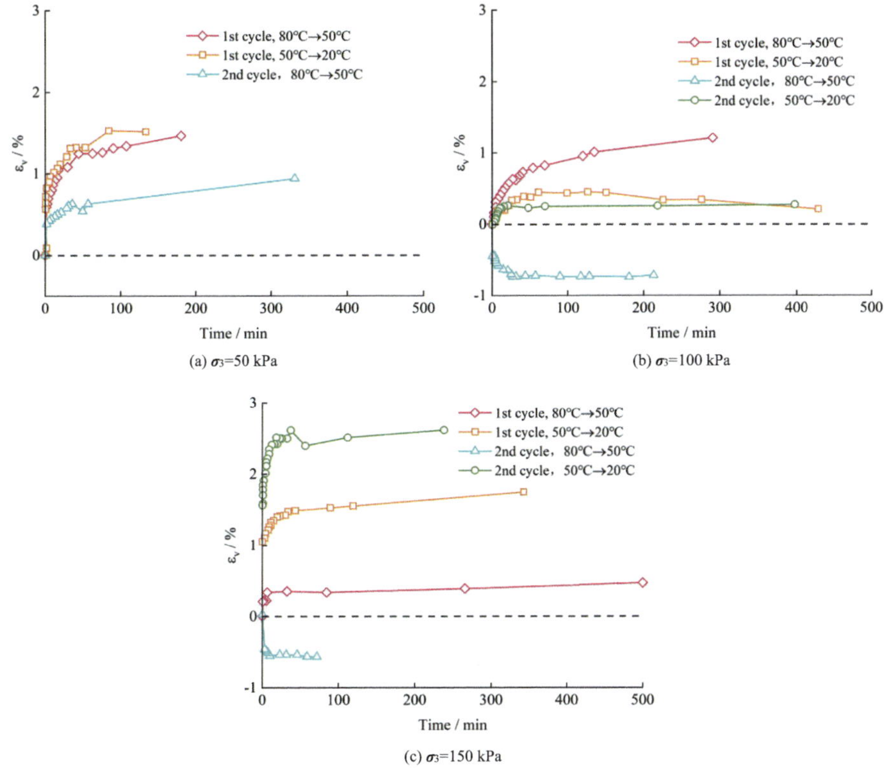

Fig. 4.16 Volumetric strain under isothermal and drained conditions after temperature dropping

pore water during the coupling process of external load (perimeter pressure) and temperature load (warming or cooling), showing the irreversibility of the thermal consolidation process of saturated soil.

(3) The specimen exhibits a consolidation drainage process under isothermal drainage conditions after undrained heating. Specimens in the undrained cooling after isothermal conditions, in addition to a few processes shown as water absorption, most of the performance of the consolidation drainage process. However, the heating process caused by the drainage body strain is much larger than the cooling process caused by the drainage body strain.

Chapter 5
Thermal Consolidation Tests on Saturated Clay Subjected to Thermal Loading on Inner and Outer Surfaces of Hollow Cylindrical Specimens

An axial thermal consolidation test apparatus suitable for saturated soils is developed. Using this apparatus, thermal loadings can be applied on both the inner and outer surfaces of a hollow cylindrical specimen separately. Thus, the specimen will be subjected to the effect of a gradient temperature along the diameter and also a lateral mechanical loading. This apparatus can be used for some special heating paths and mechanical loading paths. The consolidation of a saturated soil is accomplished, and the thermal loading on the inner and outer boundaries of the specimen is applied by steps alternately. The temperature range is 25–75 °C, and the lateral pressures are 50, 100, 150, and 200 kPa. The results show that the pore pressures and consolidation volumetric strains of the specimens induced by heating steps and various heating orders are very different even under the same state of temperature loading, which actually indicates the obvious influences of heating paths.

5.1 Introduction of the Thermal Consolidation Test Device for Hollow Cylindrical Specimens

The device used for the test is an axisymmetric hollow cylindrical specimen temperature-controlled consolidation apparatus independently developed by the tutor, as shown in Fig. 5.1.The dimensions of the hollow cylindrical specimen are: inner radius $a = 70$ mm, outer radius $b = 150$ mm, height $h = 280$ mm, i.e., the height to diameter ratio is $h/2b = 1.87$. The pressure chamber can withstand high radial consolidation pressures (pressures of up to 1 MPa) and high temperatures (temperatures are set at 0–100). The sidewall of the pressure chamber is a cavity structure made of stainless steel, with a cylindrical electric heating plate placed inside the sidewall. The temperature inside the pressure chamber is controlled by a temperature sensor installed in the pressure chamber and a temperature regulation system

G. Yang and B. Bai, *Experimental and Theoretical Study on Thermal Consolidation Behavior of Saturated Clay*, https://doi.org/10.1007/978-981-96-6909-7_5

connected to it (i.e., controlling the external boundary temperature of the specimen), with an accuracy of up to 0.1 °C and a heating power of 500 W.

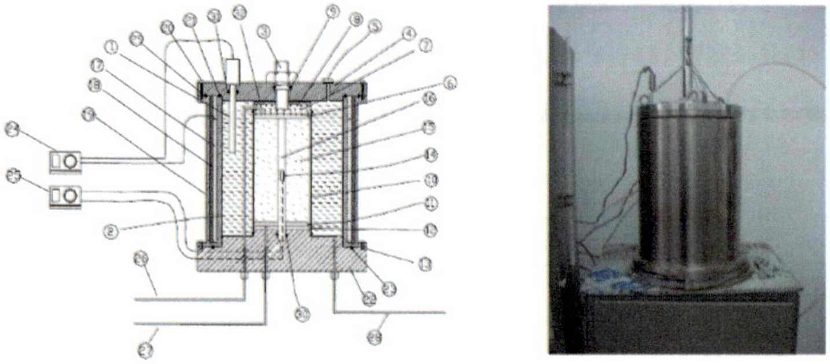

(a) Pressure chamber structure (b) Pressure chamber appearance

Fig. 5.1 Consolidation test apparatus suitable for hollow cylindrical specimen. *Note* 1 Pressure chamber electric thermocouple; 2 Pressure chamber sidewall; 3 Axial pressure rod; 4 Pressure chamber top cover; 5 Water injection hole bolt; 6 Upper permeable stone; 7 Specimen cap; 8 Adiabatic ceramic plate; 9 O-rubber gasket sealing ring; 10 Cylindrical rubber membrane; 11 Lower permeable stone; 12 Adiabatic ceramic plate; 13 Fixing bolts; 14 Temperature sensor; 15 Hollow cylindrical specimen; 16 Hot rod; 17 Electric heating plate; 18 Asbestos; 19 pressure chamber outer side wall; 20 fixing bolt; 21 O-type rubber gasket sealing ring; 22 pressure chamber base; 23 O-type rubber ring; 24 temperature regulator; 25 temperature regulator; 26 specimen drain pipe; 27 pore water pressure force measurement pipe; 28 side pressure connection pipe; 29 O-type rubber gasket sealing ring; 30 O-type rubber ring; 31 O-type rubber sealing ring; 32 water-stopping rubber gasket

Fig. 5.2 Control and monitoring devices

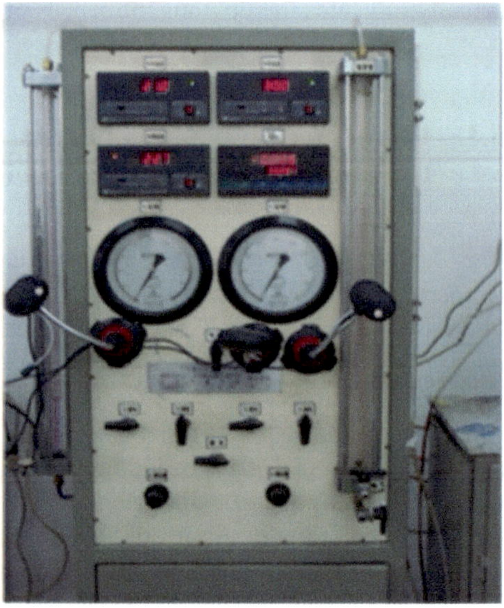

The designed hollow cylinder specimen temperature-controlled consolidation apparatus adopts modular combination structure, which is mainly composed of pressure chamber, side pressure, and counterpressure control system, water–gas-circuit control cabinet and microcomputer and data monitoring system, etc., and is accompanied by a precise peripheral pressure control system, pore water pressure force measuring element, drainage volume measuring device, and digital camera system. The system is composed of the following parts: pore water pressure control system, drainage volume measuring device and digital camera. Hollow cylindrical specimen is wrapped with a special impermeable cylindrical latex membrane, and the volume change of the pore water inside the specimen is measured by the counterpressure drainage system (double-barrel variable tube). Consolidation pressure, pore water pressure, and conventional triaxial instrument peripheral pressure control system, pore water pressure force measurement system set, to impose and automatic measurement.

Peripheral pressure connecting pipe, sample drain, pore water pressure measurement pipe, and the system connecting pipeline are made of stainless steel; the middle section is made into a spiral structure and placed into a temperature constant water container (controlled and regulated by another set of temperature control system). From the temperature, constant water container leads to the pore water pressure force measurement tube and pore water pressure force measurement system connected to the specimen drain tube and body change tube connected to the pressure around the connecting tube and the pressure regulator system connected to the pressure regulator. In this way, the specimen drainage volume can be measured at a constant temperature (such as the ambient temperature), eliminating the thermal expansion and contraction effect of water on the volume change, to ensure that the temperature changes in the process of the specimen solid body strain accurate measurement.

This test setup can be used to simulate engineering problems that can be regarded as one-dimensional cases such as hot piles or holes in geothermal development [1, 2], and it can also simplify the temperature boundary conditions of the test, which facilitates the determination of the basic parameters of thermodynamics.

5.2 Sample Preparation

The soil used in this test is the clay commonly found in the northern region. The preparation of good specimens is the basis for obtaining accurate test results, in order to ensure the reliability of the test data and can be compared and analyzed with the results of previous research, the preparation of specimens in accordance with the unified standard operation. The preparation of this specimen is similar to the preparation of soil samples for conventional triaxial tests, but due to the fact that the volume of the specimen is much larger than that of the conventional test specimen and needs to be made into a hollow cylindrical shape, specific tools are required for the preparation of the specimen. The sample preparation tools used in the test were developed independently and used in conjunction with the test device, which is easy to operate and accurate.

Through the indoor test, the basic physical property indexes of the soil samples were measured, as shown in Table 5.1.

After determining the plasticity index of this soil sample $10 < I_p < 17$, it can be determined that the test soil sample is a pulverized clayey soil according to the classification of fine-grained soil.

Specific steps for specimen preparation are as follows:

(1) Take enough soil and dry it in the oven for 8 h, break the dried soil with a pulverizer, and sieve it through a 0.5 mm sieve according to the experimental requirements;
(2) Sieve out about 7000 g of dry soil for test, and calculate the amount of water needed to prepare the specimen according to the 15% moisture content (close to the optimal moisture content) required for the test;
(3) Take 600 g of sieved dry soil into the soil mixing basin at one time, weigh the corresponding required amount of water (evaporation or spattering in the soil mixing process, weigh out a few more grams), and mix thoroughly, and pass through a 2 mm sieve, mixing in eleven times;
(4) Mix the soil mixed at 15% water content, put it into a sealed bag, and moisten it for one day and one night;
(5) Determine the moisture content of the soil sample after wetting, take different locations of the soil, not less than two points, the difference between the moisture content of the sample and the required moisture content shall not be greater than 1%;
(6) Take layered compaction method to prepare soil samples. Control of soil samples dry density must be, according to the measured preparation of wet soil measured moisture content to calculate the required amount of soil, six layers to compact soil samples, the height of each layer of 46.67 mm, and ultimately produced a height of 280 mm, the outer diameter of 150 mm, the inner diameter of the ring specimen of 70 mm. When struck in layers, scraping is required between every two layers;
(7) The specimen is saturated by vacuum pumping saturation method, as shown in Fig. 5.3. Will be placed in the saturator in the specimen into the saturated cylinder, cylinder and between the top cover coated with a thin layer of petroleum jelly cover tightly in order to prevent air leakage; the use of vacuum pump pumping, in the negative of an atmospheric pressure on the specimen for continuous pumping for 3 h; pumping is completed open the water inlet valve will be

Table 5.1 Basic physical indicators of sample

Initial moisture content (%)	Saturation (%)	Proportion	Dry density (g/cm^3)	Liquid limit (%)	Plastic limit (%)	Plasticity index (%)
19.7	98.0	2.72	1.66	29.95	16.05	13.9

airless water (distilled water or boiled water) using the saturated cylinder of the negative pressure is slowly injected into it (can be varied pumping side injection to keep the pressure unchanged), to be the water surface is not over the specimen to stop injecting water. Stop injecting water when the water surface does not exceed the specimen, and then open the valve to make the saturated cylinder connected to the atmosphere; under atmospheric pressure, let the specimen soak for 12 h so that it is fully saturated; saturated after weighing the specimen to calculate the degree of saturation, the degree of saturation shall not be less than 95%.

The formula for calculating saturation is as follows:

$$S_r = \frac{(\rho - \rho_d)G_s}{e\rho_d} \quad \text{or} \quad S_r = \frac{wG_s}{e},$$

where s_r—Degree of saturation, %; ρ—Density after saturation; ρ_d—Dry density of the soil; e—Pore ratio of soil; G_s—Specific gravity of soil particles; w—Water content after saturation, %.

Fig. 5.3 Vacuum extraction

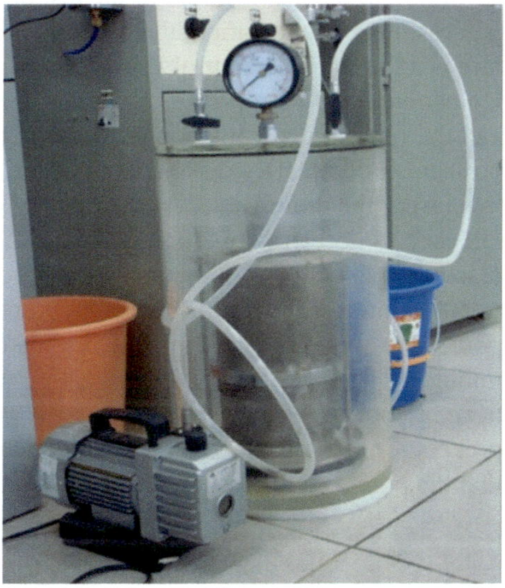

5.3 Thermal Consolidation Test of Hollow Cylindrical Specimens Under Isothermal Conditions at Internal and External Boundaries

5.3.1 Test Program

Based on the main research and development of the ring temperature-controlled consolidation instrument that can control the temperature of the inner and outer boundaries, the evolution of the pore pressure and consolidation deformation of saturated clayey soils is investigated under the conditions of different consolidation pressures (50, 100, 150, and 200 kPa) and different temperature gradients and gradients (see Table 5.2 for details).

All specimens were initially consolidated at the initial temperature (25 °C), which was the same as the conventional triaxial consolidation test, and the initial consolidation time was 1440 min, during which the pore water pressure should be completely dissipated to stabilization. After the initial consolidation is completed, the internal and external boundaries are simultaneously heated to a predetermined value according to the heating path of the test program, and the process is a non-drainage process, i.e., when heating, the drainage valve is closed first and then the temperature is adjusted. The control time of the warming process is 120 min, and after the temperature and pore pressure are stabilized, consolidation drainage is carried out, and the time of consolidation drainage is controlled to be 1440 min until the pore pressure is completely dissipated to stability. Then undrained cooling is conducted by closing the drain valve. As the cooling process relies on natural cooling (with a fan used to accelerate the cooling rate), the temperature is allowed to drop until it reaches and stabilizes at the predetermined temperature, which typically takes 300 to 500 minutes. The drain valve is then opened for consolidation drainage until the pore water pressure is completely dissipated and stabilized, a process lasting 1440 minutes. The above constitutes a heating-cooling cycle. It should be noted that the heating-cooling process for the inner and outer rings is undrained, the consolidation process is always drained consolidation, and the consolidation process must ensure that the pore water pressure is completely dissipated to stability.

5.3.2 Test Procedure

The test was conducted according to the designed test program with the following procedure:

(1) Preparation of saturated soil samples.
(2) Initial adjustment of the control instrument.

The control instrument is shown in Fig. 5.2, turn on the power, and check whether each instrument panel shows normal and whether the air compressor can provide

Table 5.2 Test plan (at room temperature 25 °C)

Heating path	Perimeter pressure (kPa)	Internal heat source temperature (°C)	External heat source temperature (°C)	Description of experimental procedures and points	Number of samples
1	50 100 150 200	25 → 40 → 25	25 → 40 → 25	Initially consolidates and drains at 25 °C ambient temperature; consolidates and drains as both internal and external temperatures increase; then consolidates and drains as it drops to 25 °C ambient temperature	4
2	50 100 150 200	25 → 50 → 25	25 → 50 → 25	Same as A1	4
3	50 100 150 200	25 → 60 → 25	25 → 60 → 25	Same as A1	4
4	50 100 150 200	25 → 75 → 25	25 → 75 → 25	Same as A1	4
5	550 100 150 200	25 → 40 → 50 → 60 → 75 → 60 → 50 → 40 → 25	25 → 40 → 50 → 60 → 75 → 60 → 50 → 40 → 25	Initial consolidation and drainage at an ambient temperature of 25 °C; simultaneous stepwise increases in internal and external temperatures, with consolidation and drainage at each stage; then stepwise decreases to an ambient temperature of 25 °C, with consolidation and drainage at each stage	4

stable pressure. Adjust the inner and outer ring temperature to 25 °C; the control method is to set the upper and lower limits of the control temperature, set the upper limit for 25.0 °C, the lower limit for 24.8 °C. Open the perimeter pressure inlet valve and counterpressure inlet valve, respectively, to the perimeter pressure measurement tube and counterpressure measurement tube into the appropriate amount of water, so that it is in the appropriate scale to facilitate the readings and recording of data. Open the perimeter pressure valve, apply a certain perimeter pressure, so that the water in the perimeter pressure tube from the pressure chamber base connectivity slowly out of the tube air, close the pressure valve to remove the perimeter pressure. Similarly, open the counterpressure pressurizing valve, apply a certain counterpressure, so that the water in the counterpressure tube flows out of the connecting port in the top cap of the specimen, discharging the air in the tube, and close the valve to remove the counterpressure. Open the drain valve, use a syringe to inject water from the hole pressure measurement hole, discharge the air in the tube, and then close the valve. In short, to ensure that the connecting pipe is fully filled with water to improve the accuracy of the test. At the same time, check the connecting port and joints for water leakage.

(3) Hole pressure calibration

Under the condition of not loading the specimen, close the pressure chamber and fill it with water. Open the control instrument, adjust the hole pressure digital display controller, and the initial hole pressure to zero. Apply a certain pressure (e.g., 100 kPa, 200 kPa) and adjust the pore pressure digital display controller so that it shows the same value as the applied pressure. Apply any pressure, see if the pore pressure display is the same as the applied pressure, and complete the pore pressure calibration.

(4) Specimen installation

1) Take out the saturated specimen from the saturated cylinder, remove the saturator, and mount it on the test bench. Slowly remove the three-valve membrane, the pre-prepared eight filter paper strips and soil samples around (the upper edge and the top surface of the filter paper connected to the bottom side to leave a distance of about 2 cm, so that it does not affect the measurement of pore pressure), and to the soil samples of moisture spraying and degassing.
2) Take the latex membrane with a diameter of 150 mm, a length of 430 mm and a thickness of 0.6 mm, and check whether it is broken or not. Then set it on the membrane tube, with the suction ear ball to draw out the latex membrane and the air in the membrane tube, so that it is close to the inner wall of the tube. Will be set in the membrane tube on the latex membrane carefully set on the sample, take away the suction ear ball so that the latex membrane attached to the sample, turn up the ends of the membrane tube, put on the sample top hat, latex membrane at both ends should be used to rubber band to prevent leakage of water.
3) Installation of the pressure chamber outside the cover, with screws to tighten it with the base, the upper part of the connection with the sample top hat into the rubber band to tighten, to prevent water leakage in all places. By the pressure chamber outer cover upper water filling hole to the pressure chamber filled with water, and then use the screws to seal the small mouth. Finally check whether there is water leakage in all places to avoid errors in the test process; the installation process is shown in Fig. 5.4.

(5) Initial drainage consolidation

Fig. 5.4 Installation process of sample

Turn on the power supply of the instrument as well as the computer video monitoring software, and adjust the camera to the appropriate position to facilitate the camera. Adjust the inner and outer ring temperatures so that the control temperature is 25 °C, after stabilization, apply pressure. Adjust the perimeter pressure control valve, set the perimeter pressure to 50 kPa (or 100, 150, 200 kPa), and open the perimeter pressure valve, so that the perimeter pressure measurement tube of water into the pressure chamber to provide the perimeter pressure, to be stabilized after the hole pressure, the beginning of the drainage consolidation. Open the counterpressure pressurizing valve so that the water in the specimen is drained into the counterpressure measuring tube, and record the data.

(6) Drainage consolidation after warming and warming without drainage

After the initial consolidation and drainage for 1440 min, the pore water pressure is basically completely dissipated to zero and remain stable, and then carry out the non-drainage warming process, the duration of which is 120 min. After the initial consolidation is completed, close the counterpressure valve, adjust the inner and outer ring temperatures to the set value, and then begin to warm up. During the warming process, the liquid level in the peripheral pressure measuring tube rises due to thermal expansion, so the position of the camera should always be adjusted to avoid missing data when the liquid level exceeds the monitoring range. After the warming temperature is stabilized, drainage consolidation begins, and the process is the same as the initial drainage consolidation.

(7) Drainage consolidation after cooling and cooling without drainage

The cooling process is natural cooling, and a fan can be used to assist in accelerating the cooling rate until the temperature drops to room temperature of 25°C, which typically takes 300 to 500 minutes. During the cooling process, the water in the confining pressure measuring tube drops due to thermal contraction, and the camera should be adjusted in a timely manner to avoid exceeding the detection range. After the temperature stabilizes, drainage consolidation is initiated until the pore water pressure is completely dissipated to zero.

(8) Sample removal

Fig. 5.5 Sample after thermal consolidation

After the whole process of ascending–descending temperature consolidation is completed, the test is finished, turn off the instrument power supply and computer, close the perimeter pressure valve and counterpressure pressure valve, remove the perimeter pressure, pump out the water in the pressure chamber, remove the outer cover of the pressure chamber, dismantle the specimen, and the specimen after consolidation is shown in Fig. 5.5.

5.3.3 Analysis of Initial Consolidation Test Results

Initial consolidation is consolidation at ambient temperature 25 °C. Four types of consolidation enclosure pressures (50, 100, 150, and 200 kPa) were set to analyze the evolution patterns of the pore water pressure and the strain of the consolidation body during the initial consolidation process as well as the temperature fluctuations of the test based on the obtained test results.

(1) Temperature fluctuation in the consolidation process

Figure 5.6 give the fluctuation of inner and outer ring temperatures during the consolidation process. It can be seen that the fluctuation changes are basically the same under different circumferential pressures, indicating that the fluctuation of temperature has nothing to do with the circumferential pressure. In fact, the temperature value is measured by the temperature change of the water inside the pressure chamber, and the fluctuation range of the temperature inside the pressure chamber is related to the accuracy of the temperature control system. Figure 5.7 gives a comparison of the temperature fluctuations in the inner and outer ring, the frequency of change is basically the same, the fluctuation amplitude of the temperature of the outer ring of 24.5–27.4 °C, the fluctuation amplitude of the temperature of the inner ring of 24.7–26.9 °C. It can be seen that the temperature of the outer ring of the fluctuation is greater, this is due to the outer ring for the cylindrical heating device, placed on the side wall of the pressure chamber, and the heated volume of water in the pressure chamber is much greater than the volume of water in the internal hot rod. When the outer ring is heated, the pressure chamber side wall is hot due to heating, after the heating stops from the side wall of the high temperature to the pressure chamber, water heat transfer occurs, and the temperature will continue to increase. When there is no heating for a long time, the room temperature is low, the temperature of the side wall of the pressure chamber is low, and heat transfer occurs from the temperature of the water inside the pressure chamber to the outside, and the temperature continues to decrease even when the minimum setting value is reached. On the contrary, the internal heat rod wrapped around the outside of the specimen, the temperature is basically certain, the volume of water inside the rod is small, and the temperature fluctuation is small.

(2) Pore water pressure dissipation process

Figure 5.8 shows the dissipation process of pore water pressure in the consolidation process when the ambient temperature (25 °C) and the surrounding pressure are 50, 100, 150, and 200 kPa, respectively. It can be seen that the dissipation curves of pore water pressure are very similar under different ambient pressures. At the beginning of consolidation, the normalized pore water pressure (perimeter pressure/pore water pressure) of most specimens is close to 1. This indicates that the specimens are highly saturated, close to fully saturated, and the applied perimeter pressure is basically borne by the pore water completely. After the beginning of consolidation, drainage consolidation, the water in the specimen is continuously excluded, and the pore water pressure gradually decreases. From the figure, it can be seen that the dissipation rate of the pore pressure is very fast in the first 60 min, and the normalized pore water pressure corresponding to 60 min is 0, 0.01, 0.07, and 0.16, respectively, which shows the trend that the larger the peripheral pressure is, the smaller the dissipation rate is. This is due to the fact that the larger the perimeter pressure is, the larger the pore pressure is reached without drainage and pressurization, and the longer the time to dissipate the pore pressure for drainage and consolidation. In the whole consolidation process, the time corresponding to the complete dissipation of pore water pressure with different perimeter pressures are 60, 100, 280, and 420 min, respectively, and the consolidation time set for the test is 1440 min, which can ensure the complete dissipation of pore pressure. The relationship between confining pressure and pore water pressure dissipation can be observed: the higher the confining pressure, the lower the dissipation rate of pore water pressure, and the longer the time required to reach stability.

(3) Volumetric strain with time change process

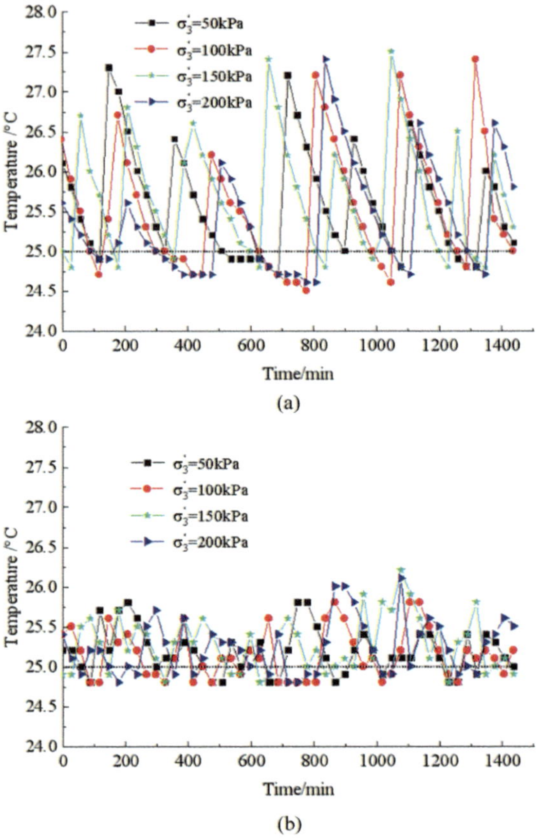

Fig. 5.6 **a** Outer ring temperature fluctuations; **b** inner ring temperature fluctuations

The strain referred to in this paper is the counterpressure body strain (the volume of water discharged from the specimen) and the corresponding peripheral pressure body strain, that is, the volume of water in and out of the pressure chamber. Figure 5.9 gives the ambient temperature (25 °C), four kinds of peripheral pressure (50 kPa, 100 kPa, 150 kPa, and 200 kPa) under the consolidation process of the specimen body strain over time. It can be seen that the greater the perimeter pressure, the greater the deformation of the specimen, the greater the volume of discharged water, the greater the body strain of the specimen, and the final body strain was 2.2%, 2.6%, 3.5%, and 4.8%, respectively. Corresponding to the change rule of pore pressure, the rate of body strain change in the first 60 min is very large, which is also the main time period of body strain change, and the body strain corresponding to different perimeter pressures at 60 min is 2.1%, 2.3%, 2.9%, 4.2%, respectively. The final time for the stabilization of the body strain change is about 400 min, which corresponds to the dissipative stabilization law of the pore pressure, and the larger the perimeter pressure, the longer the time to reach stabilization.

Fig. 5.7 **a** Outer and inner ring temperature fluctuations at 50 kPa; **b** outer and inner ring temperature fluctuations at 200 kPa

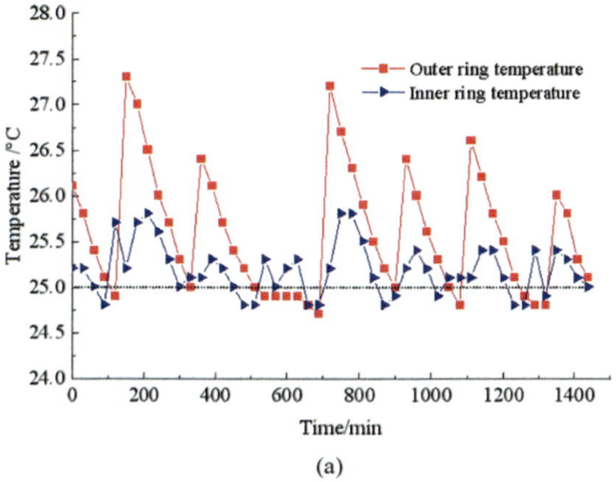

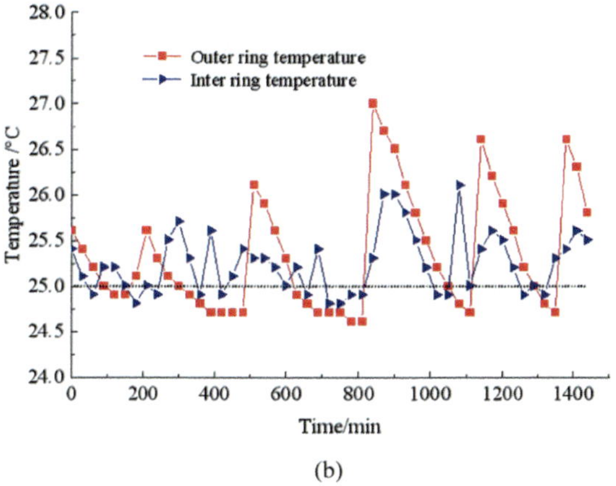

Fig. 5.8 Pore water pressure dissipation process during initial consolidation

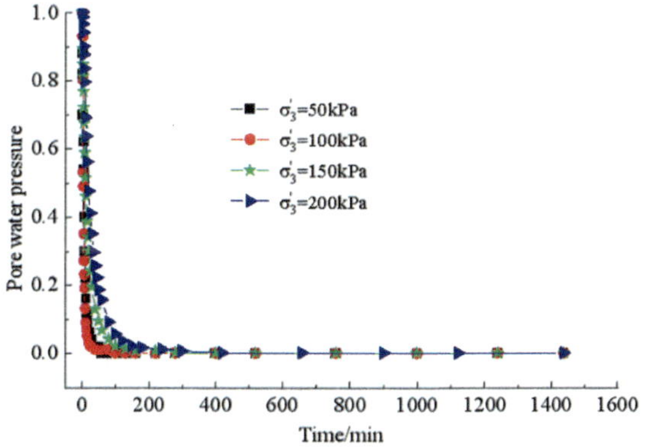

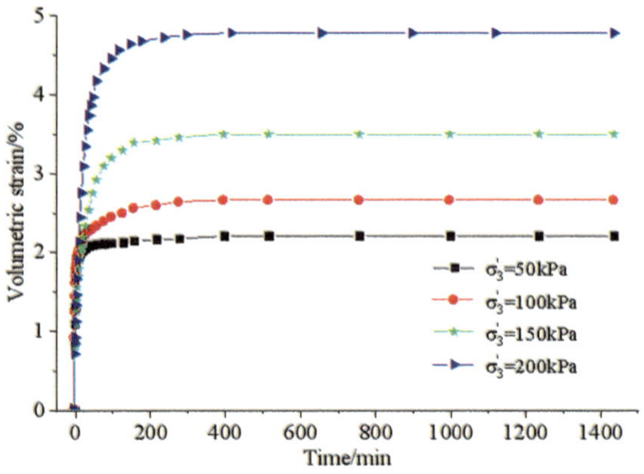

Fig. 5.9 Volumetric strain change during initial consolidation

Figure 5.10 gives the relationship between the strain of the counterpressure body and the change of the strain of the enclosing pressure body with time for four enclosing pressures at ambient temperature (25 °C). It can be seen that the two change curves are basically the same. That is, since the soil particles and pore water inside the specimen have almost no expansion or contraction effect at room temperature, the volume change of the specimen is equal to the discharge of pore water inside the specimen. Since the volume change of the specimen, the water inside the perimeter pressure tube enters the pressure chamber, and the amount of pore water discharged is basically the same as the amount of pore water discharged out, so the volume change of the specimen at room temperature can also be measured by the change in the volume of water inside the pressure chamber. Peripressure body strain in the solidification and stabilization there is a small up and down ups and downs; this is due to the fluctuation of the test temperature caused by the contraction or expansion of the volume of water inside the pressure chamber.

5.3.4 *Thermal Consolidation Response Due to Single-Stage Heating–Cooling Process*

The so-called single-stage ascending–descending temperature consolidation process is the path given by groups 1 to 4 in Table 5.2 of the test program (25 → 40 → 25 °C, 25 → 50 → 25 °C, 25 → 60 → 25 °C, 25 → 75 → 25 °C). The ascending and descending process temperatures are undrained and the consolidation process is drained.

Undrained condition. The assumption of undrained conditions for ascending and descending temperatures of saturated soils is considered valid. Based on this, the superporous water pressure arises due to: (1) the difference in thermal expansion of soil solid particles and pore water, and (2) physicochemical effects. Pore water

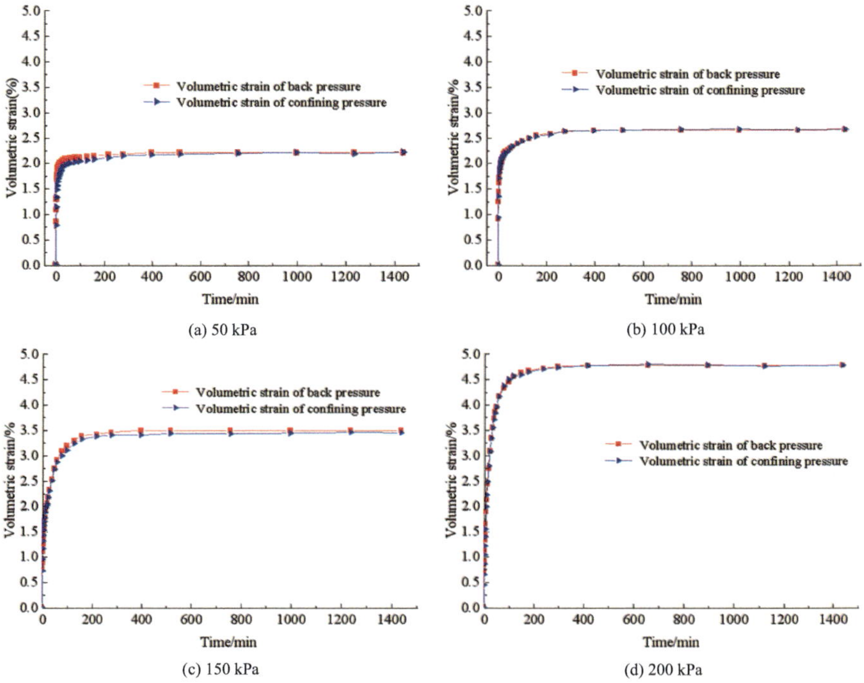

Fig. 5.10 Volumetric strain of back pressure and confining pressure

pressure caused by warming and cooling is an important factor in thermal consolidation modeling. The amount of increase in pore water pressure from undrained warming and cooling is related to the path of warming and cooling, the incremental increase in temperature, the boundary conditions, the compressibility of the soil, and the thermal expansion properties of the pore water and the soil. It is difficult to calculate this increase in pore water pressure based on theoretical foundations alone.

Drainage conditions. The dissipation of temperature-induced superporous water pressures under drained conditions and the accompanying change in volume is called "thermal main consolidation." A component of compression that is not caused by the superporous water pressure, but is still in progress when the superporous water pressure is too small to be measured, is called "thermal secondary consolidation." If the increase in temperature under fully drained conditions is small, the increase in pore pressure will also be small. When the temperature increases under fully drained conditions, it is difficult to distinguish between thermal primary consolidation and thermal secondary consolidation. Thermal secondary consolidation is related to the physicochemical effects of increased temperature.

In addition to primary and secondary consolidation, the relative thermal expansion of the soil constituents causes water to leave the soil skeleton, and this volume change is completely elastic. The change in pore volume controls only the mechanical properties of the soil. Therefore, the water discharged measured during heating needs to be adjusted by the thermal expansion treatment. Three processes of thermal consolidation can therefore be assumed, primary consolidation, secondary consolidation, and thermoelastic expansion, and these phenomena can be modeled to predict the properties of saturated soils when heated.

(1) Temperature change

Figure 5.11 gives the temperature changes of inner and outer rings in the test program for 1–4 groups with different amplitudes of temperature rise and fall. In general, the temperature changes of inner and outer rings are similar in different amplitudes of temperature rise and fall. Figure 5.12b gives the temperature change of the lifting temperature path 25 → 60 → 25 °C under different enclosure pressures, and it can be seen that the change curves under different enclosure pressures are almost the same, which proves that the temperature change of the lifting temperature process does not have much relationship with the enclosure pressure.

Based on this, Fig. 5.11 only gives the temperature change of the lift temperature under the enclosure pressure of 100 kPa. In the primary stage of the undrained heating process, the temperature rise rate is large, and with the gradual increase in specimen temperature, the temperature rise rate gradually slows down. The whole heating process lasts about 100–120 min, and the temperature rises from 25 °C to the

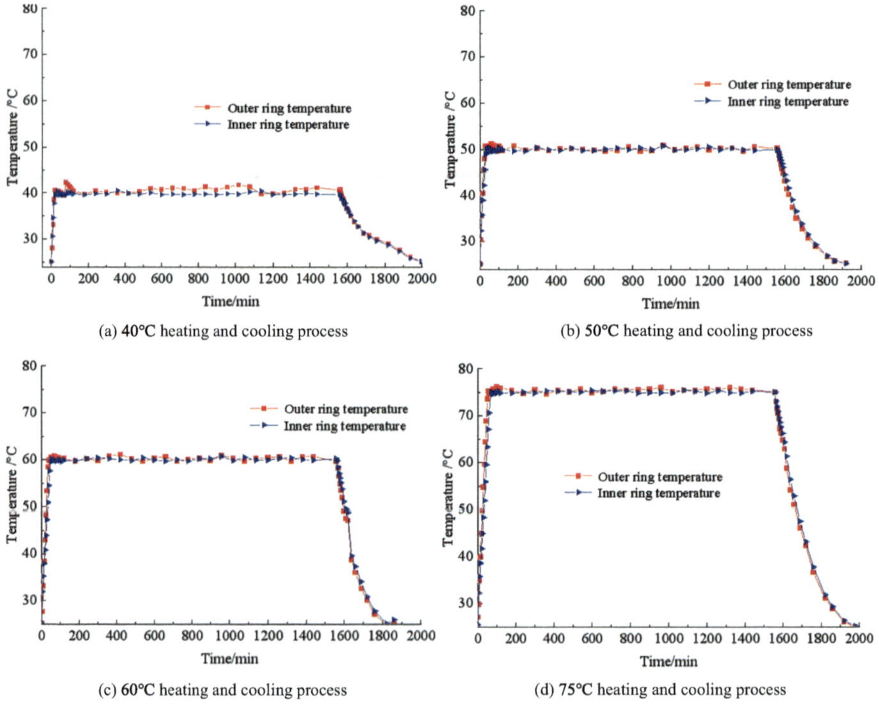

Fig. 5.11 Temperature changes during heating–cooling consolidation

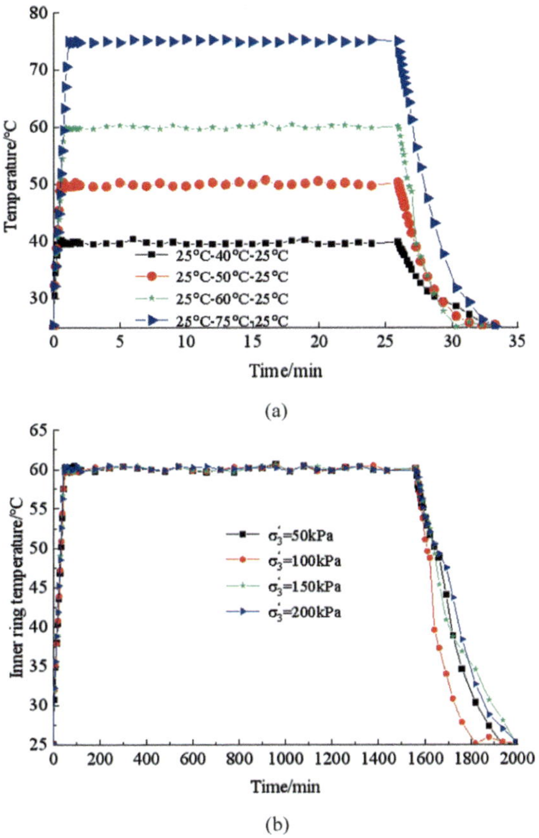

Fig. 5.12 **a** Temperature changes of different heating and cooling amplitude; **b** comparison of temperature changes at different confining pressure

set values (40, 50, 60, 75 °C) and stabilizes. Subsequently, drainage consolidation, which is maintained at a certain temperature, begins and lasts for 1440 min, with the temperature fluctuating within the range of − 0.5 ~ 1 °C at this stage. After the consolidation was completed, the cooling was started without drainage, and since the process was natural cooling (room temperature was about 20 °C), the temperature drop rate was initially large, and with the gradual decrease of temperature, the temperature drop rate was gradually slowed down. The whole cooling process lasts about 300–500 min until it drops to 25 °C and stabilizes. Finally, the solidification and drainage at 25 °C, which is not given in the figure, is the same as the fluctuation of the set value of 25 °C above, and the duration is also 1440 min.

In Figure 5.12a, grouping the four temperature paths into the same coordinate graph, it can be seen that the heating rate is basically the same for different heating amplitudes, while the cooling rate is slightly different due to the natural cooling influenced by the room temperature. At the initial stage of drainage consolidation, i.e., after 120 min, the body strain decreases due to the outflow of pore water, and the water at room temperature flows into the pressure chamber more, and the temperature fluctuation is more frequent under its influence. After that, as the consolidation process proceeds, the exchange of water between inside and outside the pressure chamber becomes less, and the temperature is relatively stable. The reasons for the temperature fluctuations are summarized as the accuracy of the instrument (± 1 °C), the change of temperature in the chamber, the heat exchange between the pressure chamber and the environment, and the inflow and outflow of water in the pressure chamber at the initial stage of consolidation and drainage.

(2) Evolution of Pore Water Pressure

Figure 5.13 shows the normalized pore water pressure (pore water pressure/ambient pressure) during the undrained heating-drained consolidation-undrained cooling process for four different temperature change paths, with confining pressures σ_3 of 50, 100, 150, and 200 kPa. Overall, it can be observed that under the same temperature path, the variation in pore water pressure is consistent across different temperature amplitudes. The larger the confining pressure, the smaller the absolute value of the maximum normalized pore water pressure achieved during the heating and cooling process. For the same confining pressure, a larger temperature amplitude results in a greater absolute value of the maximum normalized pore water pressure, up to the potential maximum value of 1. Under the same temperature load, as the confining pressure increases, the rate of pore water pressure increase slows down. This can be attributed to the fact that the sample was previously subjected to significant consolidation pressure, leading to a noticeable reduction in pore water within the sample.

For the 40 °C heating and cooling process (Fig. 5.13a), as the temperature increases, the pore pressure rises. When the temperature stabilizes at 40 °C, the pore pressure also reaches its maximum value, which takes approximately 100 min. The maximum normalized pore water pressures achieved during heating for the four confining pressures (50, 100, 150, 200 kPa) are 0.46, 0.27, 0.2, and 0.185, respectively. As consolidation drainage proceeds, the pore pressure begins to dissipate. Initially, this dissipation is rapid, reaching a very small value within 100 min, and then it slowly dissipates to zero and stabilizes. Undrained cooling is then conducted. Due to the fact that the volumetric deformation of pore water contraction is significantly greater than that of soil particles (the thermal expansion coefficient of water is approximately 2×10^{-4} °C^{-1}, whereas the thermal expansion coefficient of soil particles is generally 1.5×10^{-5} °C^{-1}), negative pore water pressures are generated, resulting in an irreversible pore water pressure change process. Furthermore, as the confining pressure increases, the magnitude of the negative pore water pressure also increases. The maximum normalized negative pore water pressures achieved during cooling for the four confining pressures are – 0.2, – 0.12, – 0.08, and – 0.06. When the temperature of the sample drops to 25 °C and the drainage valve is opened, due to the presence of negative pore water pressure, the sample undergoes a water absorption process, and the pore water pressure gradually rises and approaches zero. Similarly, the variation patterns during the 50, 60, and 75 °C heating and cooling processes are

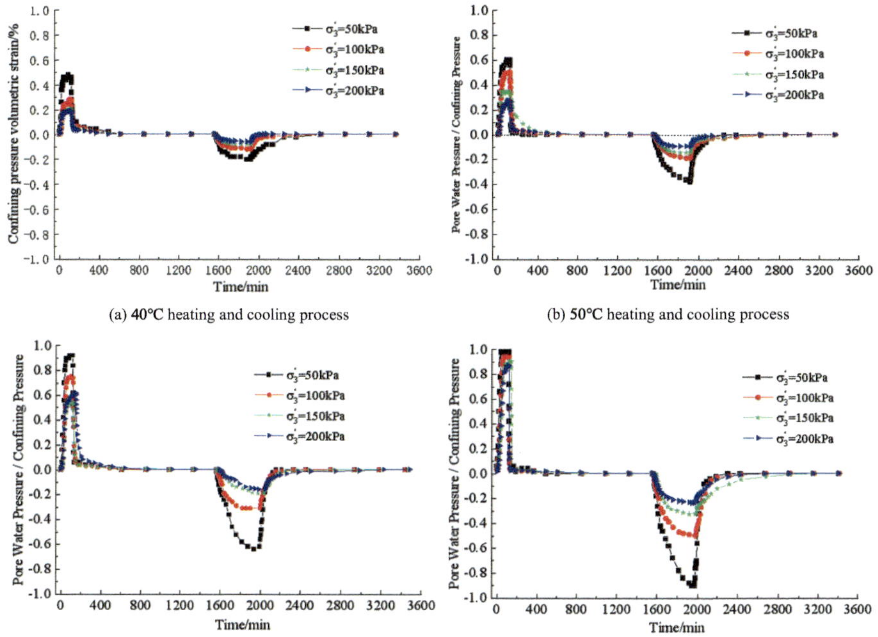

(a) 40°C heating and cooling process
(b) 50°C heating and cooling process
(c) 60°C heating and cooling process
(d) 75°C heating and cooling process

Fig. 5.13 Normalized pore water pressure evolution process

similar to those at 40 °C and will not be discussed in detail. The maximum normalized pore water pressure values achieved during heating and cooling are listed in Table 5.3.

Figure 5.14 illustrates the comparison of normalized pore water pressure and temperature evolution during the undrained heating and cooling process at a confining pressure of 100 kPa and a temperature of 60 °C. This figure provides a more intuitive representation of the relationship between temperature and pore pressure during undrained heating and drained consolidation. During undrained heating, due to the differences in thermal expansion between the solid soil particles and pore water, as well as some physical effects, the water binding around certain soil particles is activated, resulting in an increase in pore water pressure as the temperature rises. Once the temperature stabilizes, the pore water pressure also reaches a stable state. During drained consolidation, the pore pressure dissipates rapidly, reaching stability in a very short time, which is slightly faster than the initial consolidation phase. After the pore pressure fully dissipates, undrained cooling begins. Since the cooling contraction of solid particles is less than that of pore water, negative pore water pressure is generated within the sample. Once the temperature drops to the ambient temperature of 25 °C and stabilizes, drained consolidation occurs, and the negative pore pressure gradually dissipates. The rate of dissipation is initially fast and then

Table 5.3 Normalized pore water pressure and volumetric strain change caused by heating and cooling

Confining pressure (kPa)	Determination of physical quantities	Temperature change amplitude			
		25 → 40 → 25 °C	25 → 50 → 25 °C	25 → 60 → 25 °C	25 → 75 → 25 °C
50	Heating $(u/\sigma_3)_{peak}$	0.48	0.60	0.90	0.98
	Cooling $(u/\sigma_3)_{peak}$	− 0.20	− 0.38	− 0.62	− 0.90
	Heating consolidation ε_{vmax}/%	0.51	0.67	1.21	2.12
	Cooling consolidation ε_{vmax}/%	0.09	0.24	0.68	1.48
	Overall strain ε_v/%	0.42	0.43	0.53	0.64
100	Heating $(u/\sigma_3)_{peak}$	0.30	0.50	0.74	0.94
	Cooling $(u/\sigma_3)_{peak}$	− 0.12	− 0.19	− 0.31	− 0.50
	Heating consolidation ε_{vmax}/%	0.54	0.93	1.31	2.15
	Cooling consolidation ε_{vmax}/%	0.06	0.18	0.66	1.48
	Overall strain ε_v/%	0.48	0.75	0.65	0.67
150	Heating $(u/\sigma_3)_{peak}$	0.21	0.34	0.59	0.90
	Cooling $(u/\sigma_3)_{peak}$	− 0.08	− 0.15	− 0.19	− 0.33
	Heating consolidation ε_{vmax}/%	0.53	0.78	1.45	2.43
	Cooling consolidation ε_{vmax}/%	0.1	0.16	0.90	1.42
	Overall strain ε_v/%	0.43	0.62	0.46	1.01
200	Heating $(u/\sigma_3)_{peak}$	0.19	0.28	0.57	0.87
	Cooling $(u/\sigma_3)_{peak}$	− 0.06	− 0.095	− 0.16	− 0.235
	Heating consolidation ε_{vmax}/%	0.61	0.81	1.31	2.26
	Cooling consolidation ε_{vmax}/%	0.09	0.18	0.30	0.95
	Overall strain ε_v/%	0.52	0.63	1.01	1.31

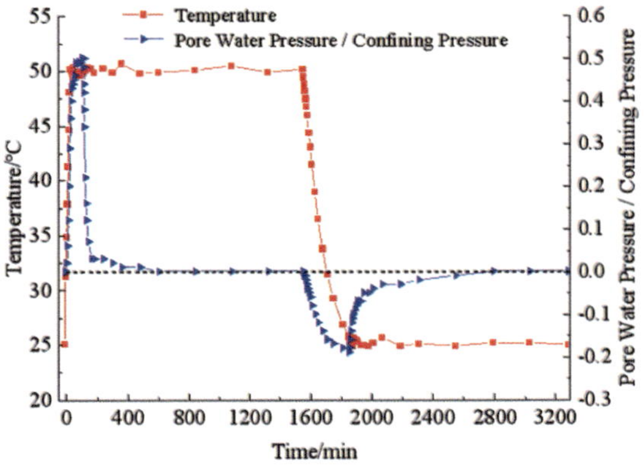

Fig. 5.14 Normalized pore water pressure and temperature evolution process during 60 °C heating and cooling consolidation process at 100 kPa

slows down, with the overall rate of negative pore pressure dissipation being slower than that of positive pore pressure, ultimately dissipating completely to 0.

(3) Consolidation Strain Evolution

Here, the term "volumetric strain" refers to the pore pressure-induced volumetric strain of a sample caused by temperature, defined as $\varepsilon_v = \Delta v/v_0$ (%, where V_0 is the initial volume of the sample, and Δv is the volumetric strain induced by temperature leading to consolidation drainage). Under drainage conditions, the pore water pressure induced by temperature dissipates rapidly and quickly stabilizes a process faster than the dissipation of pore water pressure in the initial consolidation phase. During heating, the consolidated drainage leads to a rapid increase in volumetric strain of saturated soil samples, which quickly stabilizes. During cooling, the consolidation drainage causes a negative pore pressure, resulting in a phenomenon of water absorption, leading to a decrease in volumetric strain compared to before, but overall, the volumetric strain increases.

Figure 5.15 illustrates the consolidation strain variations under four different confining pressures for four single-step heating and cooling paths. Overall, for the same temperature path, the size of consolidation strain is independent of the confining pressure [3]. Under the same confining pressure, the larger the heating and cooling amplitude, the larger the consolidation strain. The former is due to the larger consolidation pressure, leading to a greater volume of water expelled during the initial isothermal (25 °C) consolidation phase, resulting in a smaller normalized pore water pressure due to the subsequent temperature loading effect, which does not necessarily imply larger volumetric strain. The latter occurs because a larger heating and cooling amplitude creates larger pore pressures, activating more bound water around soil particles and leading to more drainage.

Analyzing the volumetric strain variation along a 75 °C heating and cooling path (Fig. 5.15d), the first 120 min represent the non-drainage heating phase where the volumetric strain is defined as 0. After heating stabilizes, drainage consolidation begins, and upon opening the drainage valve, the water in the reverse pressure

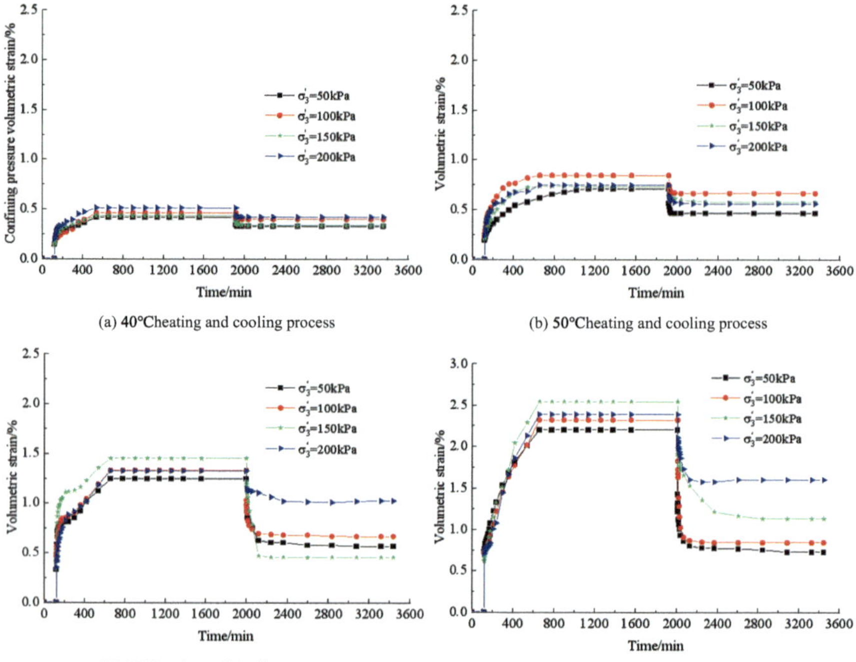

(a) 40°Cheating and cooling process

(b) 50°Cheating and cooling process

(c) 60°Cheating and cooling process

(d) 75°Cheating and cooling process

Fig. 5.15 Volumetric strain evolution process

pipe quickly rises due to the high pore pressure. Subsequently, as the pore pressure decreases, the drainage rate gradually decreases. At around 800 min of drainage consolidation, the volumetric strain stabilizes at 2.12% (50 kPa), 2.15% (100 kPa), 2.43% (150 kPa), and 2.26% (200 kPa). This process lasts longer than the dissipation of pore pressure to stability (approximately 100 min) because after the complete dissipation of pore pressure, the consolidation adjusts the internal structure of the soil, resulting in the discharge of some pore water, which is a relatively slower process. Subsequent non-drainage cooling maintains the volumetric strain. After cooling stabilizes, drainage consolidation occurs, leading to the production of negative pore pressure and water absorption phenomenon, where water from the reverse pressure pipe flows back into the sample. This process initially involves a rapid return of some water, which gradually decreases as the negative pore pressure dissipates until stability is reached at around 600 min, resulting in volumetric strains of 1.48% (50 kPa), 1.48% (100 kPa), 1.42% (150 kPa), and 0.95% (200 kPa). This process lasts longer than the dissipation of negative pore pressure to stability (approximately 300 min) due to the consolidation-induced internal structure adjustments, further changes in pore water under the influence of confining pressure. Overall, the consolidation strain increases, with final volumetric strains being 0.646% (50 kPa), 0.67% (100 kPa), 1.01% (150 kPa), and 1.31% (200 kPa), demonstrating an irreversible process of heating and cooling. The patterns of consolidation strain at 40, 50, and

60 °C are similar to those observed at 75 °C and are not further discussed here. Specific numerical values of consolidation strain can be found in Table 5.3.

Figure 5.16 shows the changes in confining pressure-induced volumetric strain under four different temperature paths and confining pressures, compared to the previously discussed back-pressure volumetric strain. Negative volumetric strain represents the volume of water expelled from the pressure chamber, while positive volumetric strain represents the volume of water entering the pressure chamber. Overall, for the same temperature path, the magnitude of the confining pressure-induced volumetric strain is independent of the confining pressure size; for the same confining pressure, the temperature variation does not affect the final volumetric strain. It is also evident that the confining pressure-induced volumetric strain is significantly influenced by temperature fluctuations after heating and lacks regularity, making it an unreliable measure for volumetric strain.

Taking the 50 °C temperature path as an example for brief analysis: Initially, without draining, heating causes thermal expansion, resulting in water being expelled from the pressure chamber. The amount of water expelled is independent of the confining pressure and corresponds to the temperature increase, with a rapid initial rate that stabilizes once the temperature reaches the set value, although with significant fluctuations. During drainage consolidation, the expulsion of pore water causes water from the confining pressure pipe to flow into the pressure chamber. Temperature

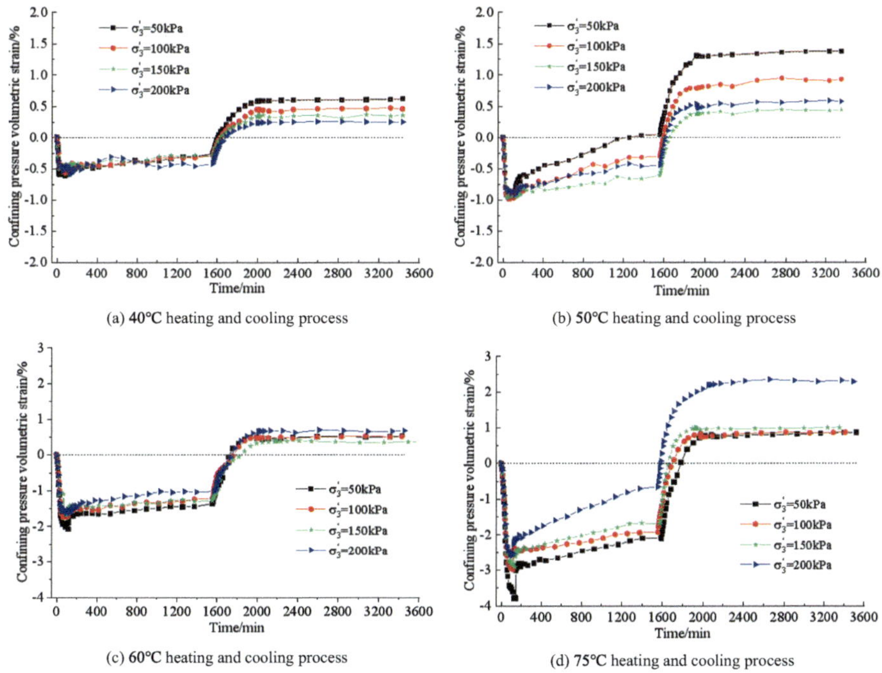

(a) 40°C heating and cooling process
(b) 50°C heating and cooling process
(c) 60°C heating and cooling process
(d) 75°C heating and cooling process

Fig. 5.16 Confining pressure volumetric strain evolution process differences between inside and outside induce minor temperature fluctuations. As consolidation stabilizes, the confining pressure-induced volumetric strain also stabilizes, but fluctuations in the external temperature can cause water to exit the pressure chamber or flow into it from the confining pressure pipe, generally leading to water entering the pressure chamber. Subsequently, cooling without drainage causes cold shrinkage, leading to continuous water inflow from the confining pressure pipe into the pressure chamber, independent of the confining pressure size. The inflow rate is initially rapid, stabilizing as the temperature lowers, followed by drainage consolidation. The confining pressure-induced volumetric strain remains mostly stable, with minor fluctuations due to temperature changes. Ultimately, the confining pressure-induced volumetric strain is positive, indicating irreversible shrinkage of the sample, but differs significantly from the back-pressure volumetric strain.

Figure 5.17 presents a comparison between the back-pressure volumetric strain and the confining pressure volumetric strain under four different confining pressures and temperature paths. Overall, it can be observed that the final volumetric strains of the two types differ; in some cases, the back-pressure volumetric strain is greater than the confining pressure volumetric strain (as shown in Fig. 5.17c), while in other instances, the confining pressure volumetric strain tends to be greater than the back-pressure volumetric strain (as seen in Fig. 5.17a, b, and d). The confining pressure volumetric strain exhibits a larger variation largely influenced by thermal expansion and contraction. The specific patterns of these changes have been analyzed in detail earlier. Due to the significant impact of temperature on the confining pressure volumetric strain, it is not suitable for accurately measuring the sample's volumetric strain; therefore, the back-pressure volumetric strain is used as the measure of the sample's volumetric strain.

(4) Relationship Between Pore Pressure, Volumetric Strain, and Confining Pressure

Table 5.3 provides specific values for the maximum normalized pore water pressure and consolidation volumetric strain achieved under single-stage temperature elevation. Based on this data, the relationship between pore pressure, volumetric strain, and confining pressure can be described.

Figure 5.18a shows that at a constant confining pressure, as the temperature increases, the normalized pore water pressure generated by non-drained heating also increases. When the temperature reaches a certain level, the normalized pore water pressure can reach a maximum of 1, meaning the pore water pressure equals the confining pressure. The change in pore pressure with temperature can be approximated as linear, and it can be observed that a higher confining pressure results in a smaller slope of this linear change. At confining pressures of 50 and 200 kPa, the linearity is not very pronounced, showing a concave-convex trend, whereas at 100 and 150 kPa, there is a clear linear change. From Fig. 5.18b, it can be seen that at the same temperature, a higher confining pressure corresponds to a lower normalized pore water pressure, and pore pressure exhibits a linear relationship with confining pressure. At a temperature of 75 °C, the change is relatively gentle; at 60 °C and 50 °C, the linear slope increases, while at 40 °C, the slope slightly decreases.

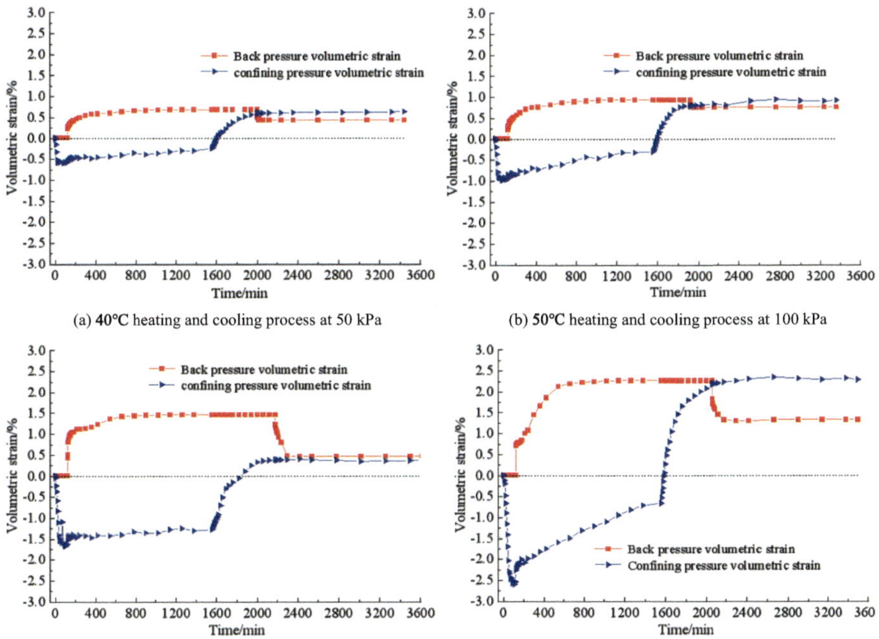

(a) 40°C heating and cooling process at 50 kPa

(b) 50°C heating and cooling process at 100 kPa

(c) 60°C heating and cooling process at 150 kPa

(d) 75°C heating and cooling process at 200 kPa

Fig. 5.17 Comparison between confining pressure and back-pressure volumetric strain

Figure 5.19 shows the variation of pore pressure with temperature during cooling. In Fig. 5.19a, it can be observed that with constant confining pressure, a larger cooling magnitude results in a higher normalized pore water pressure, which does not exceed 1. Pore pressure exhibits a clear linear relationship with temperature, with a smaller slope at higher confining pressures. In Fig. 5.19b, at the same cooling temperature, a higher confining pressure leads to a lower normalized pore water pressure. Pore pressure and confining pressure follow a parabolic relationship, with this effect being more pronounced at higher temperatures and linear at lower temperatures.

Figure 5.20 illustrates the changes in volumetric strain with temperature increase. In Fig. 5.20a, it can be observed that with constant confining pressure, volumetric strain increases with rising temperature. The rate of increase is relatively small before 50 °C but accelerates and exhibits a linear growth after 50 °C. Figure 5.20b shows that the magnitude of volumetric strain is independent of the confining pressure, and it is not possible to specifically define the relationship between volumetric strain, confining pressure, and temperature.

Figure 5.21 shows the change in volumetric strain during cooling. As mentioned, with constant confining pressure, higher temperatures result in greater volumetric strain. Overall, as temperature increases, the growth rate also increases, and the volumetric strain during cooling is independent of the confining pressure. Figure 5.21b

Fig. 5.18 Relationship between pore water pressure and confining pressure during heating process

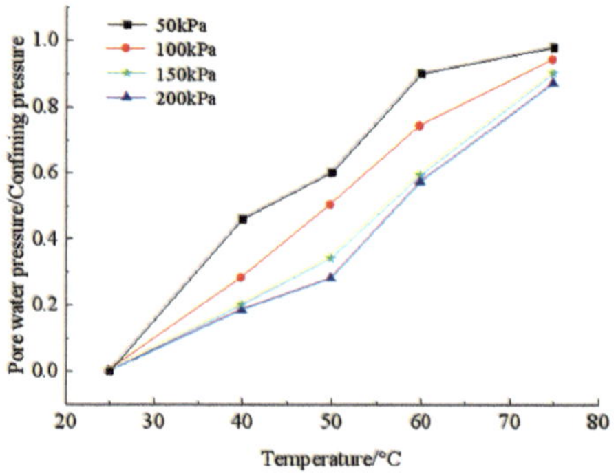

(a) Constant confining pressure, pore water pressure along with the change of temperature

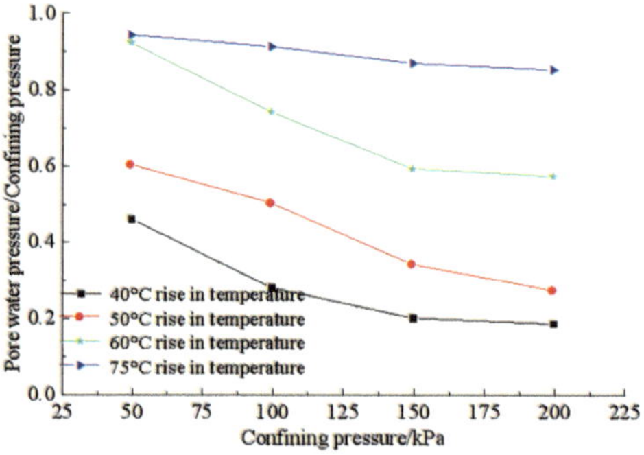

(b) Constant temperature, pore water pressure along with the change of the confining pressure

indicates that, with constant cooling, the volumetric strain decreases as the confining pressure increases.

Figure 5.22 shows the change in total strain throughout the entire heating and cooling process. It can be observed that the volumetric strain changes do not follow a regular pattern. Overall, with constant confining pressure, higher temperatures generally result in greater volumetric strain, although there are exceptions (for instance, at confining pressures of 100 and 150 kPa, the volumetric strain at 60 °C is smaller than at 50 °C). Based on the variations shown in Fig. 5.22, the change in total strain varies under different confining pressures, and the magnitude of the change is not strongly related to the confining pressure.

Fig. 5.19 Relationship between pore water pressure and confining pressure during cooling process

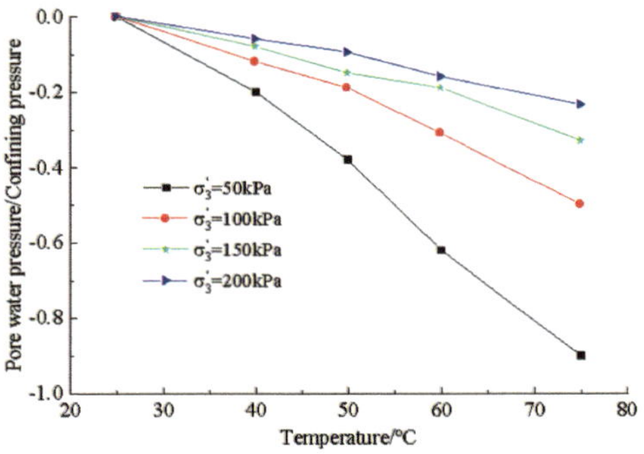

(a) Constant confining pressure, pore water pressure along with the change of temperature

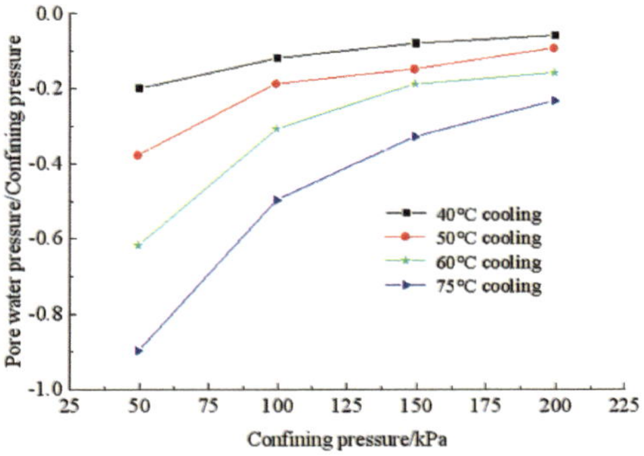

(b) Constant temperature, pore water pressure along with the change of the confining pressure

5.3.5 *Thermal Consolidation Responses Induced by Multi-stage Heating–Cooling Processes*

(1) Evolution of Pore Water Pressure

Figure 5.23 illustrates the normalized pore water pressure changes during multi-stage isothermal heating and cooling processes at inner and outer boundaries. The pattern is consistent with that observed in single-stage heating and cooling processes. Higher confining pressures result in lower maximum normalized pore water pressures during heating and cooling. For a confining pressure of 50 kPa, the maximum normalized pore water pressure during heating to 40 °C is 0.52 (the first peak), which rapidly dissipates with drainage consolidation, similar to the single-stage heating process. After 1440 min of drainage, the pressure fully dissipates. Upon increasing the temperature to 50 °C without drainage, the maximum normalized pore water pressure is 0.32 (the second peak). With further drainage and heating to 60 °C, the normalized pore water pressure reaches the third peak at 0.34. After additional drainage and heating to 75 °C, the pressure peaks at 0.56 (the fourth peak). Following drainage

Fig. 5.20 Relationship between volumetric strain and confining pressure during heating process

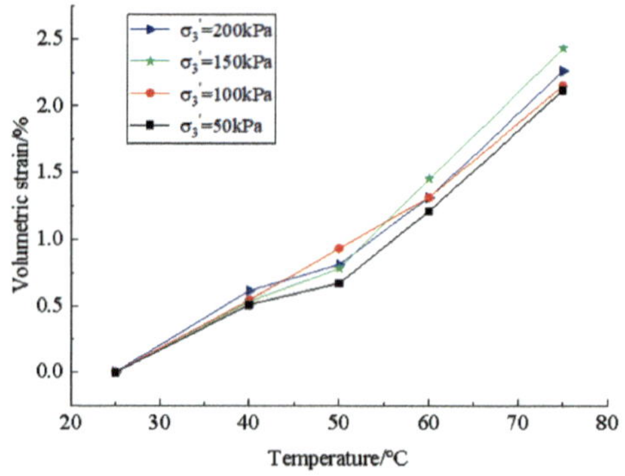

(a) Constant confining pressure, volumetric strain along with the change of temperature

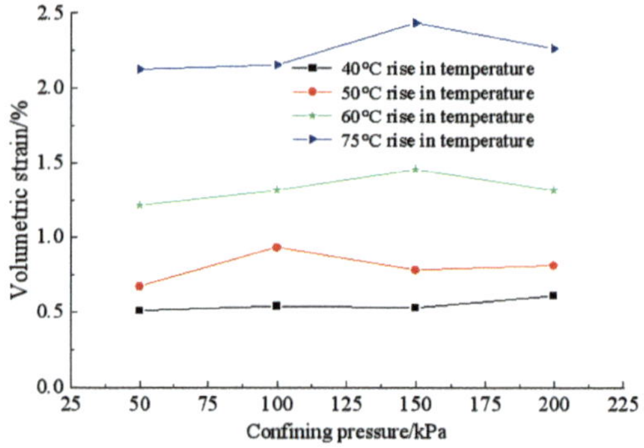

(b) Constant temperature, volumetric strain along with the change of the confining pressure

and cooling to 60 °C without drainage, the normalized pore water pressure drops to – 0.4. After further drainage and cooling to 50 °C, the pressure is – 0.24, and to 40 °C, it is – 0.22. Cooling to ambient temperature of 25 °C yields a normalized pore water pressure of – 0.26, with complete pressure dissipation after drainage. The process is similar for other confining pressures (100, 150, 200 kPa), with specific maximum normalized pore water pressure values listed in Table 5.4.

(2) Evolution of Consolidated Volumetric Strain

Figure 5.24 shows the changes in volumetric strain during multi-stage isothermal heating and cooling processes at inner and outer boundaries. It is evident that the magnitude of volumetric strain is independent of confining pressure. For a confining pressure of 100 kPa, heating to 40 °C with consolidation and drainage results in a stable volumetric strain of 0.41%. Heating further to 50 °C yields a net volumetric strain of 0.49%, with a cumulative strain of 0.90%. Heating to 60 °C results in a net strain of 0.65% and a cumulative strain of 1.55%. Finally, heating to 75 °C reaches

Fig. 5.21 Relationship between volumetric strain and confining pressure during cooling process

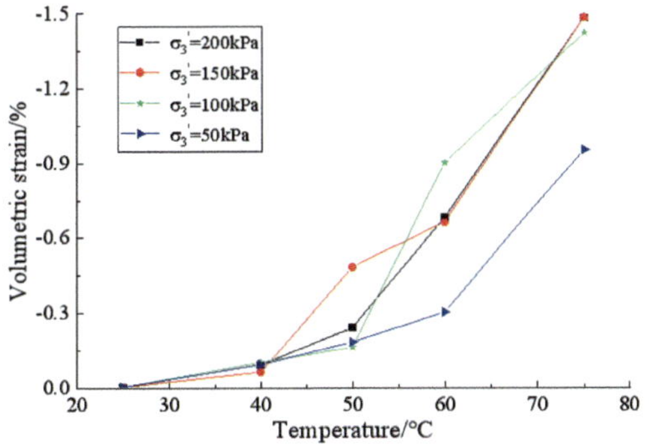

(a) Constant confining pressure, volumetric strain along with the change of temperature

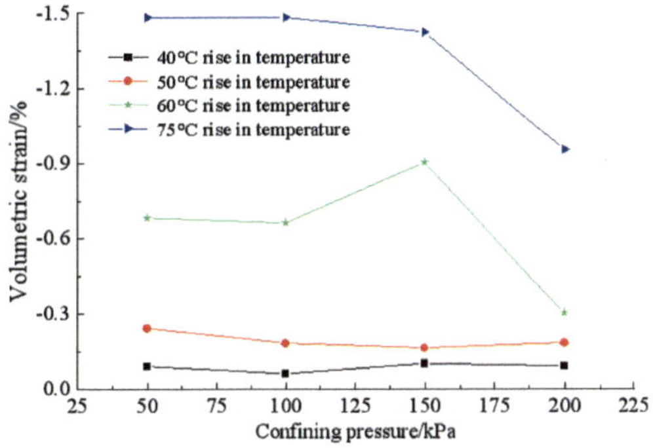

(b) Constant temperature, volumetric strain along with the change of the confining pressure

a net strain of 1.02% and a maximum cumulative strain of 2.57%. Cooling step-by-step, first to 60 °C, with consolidation and drainage, yields a net strain of − 0.51% and a cumulative strain of 2.06%. Cooling further to 50 °C results in a net strain of − 0.47% and a cumulative strain of 1.59%. At 40 °C, the net strain is − 0.48% with a cumulative strain of 1.11%. Cooling to ambient temperature (25 °C) results in a net strain of − 0.49% and a final cumulative strain of 0.62%. Similar trends are observed for other confining pressures (50, 150, 200 kPa), with specific volumetric strain values detailed in Table 5.4.

Figure 5.25 shows the variation of confining pressure strain during the isothermal multi-stage heating cooling process of the inner and outer boundaries, similar to the variation law of single-stage heating and cooling. The final strain is also positive, indicating the shrinkage of the specimen. It fluctuates greatly under the influence of temperature and differs greatly from the strain value of the back-pressure body. It can be used to measure the change in water volume in the pressure chamber during

Fig. 5.22 Relationship between volumetric strain and confining pressure during whole process

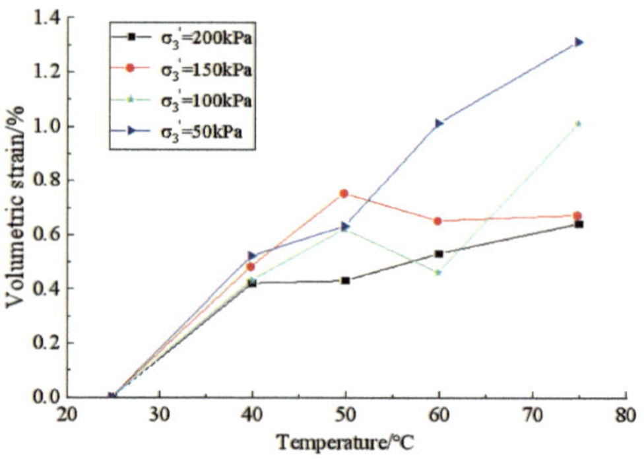

(a) Constant confining pressure, volumetric strain along with the change of temperature

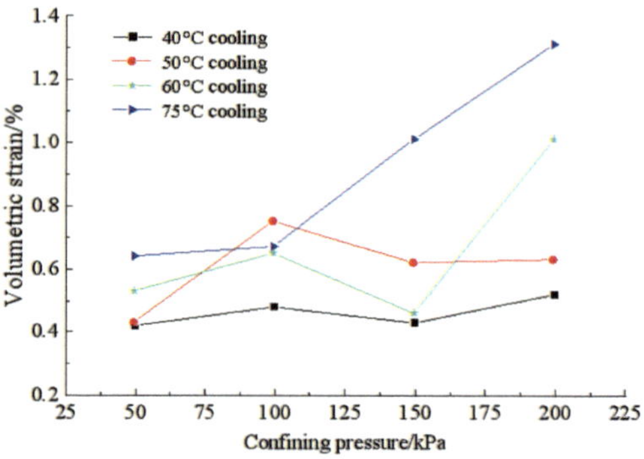

(b) Constant temperature, volumetric strain along with the change of the confining pressure

Fig. 5.23 The Normalized pore water pressure change during inside and outside boundary isothermal multi-stage heating–cooling process

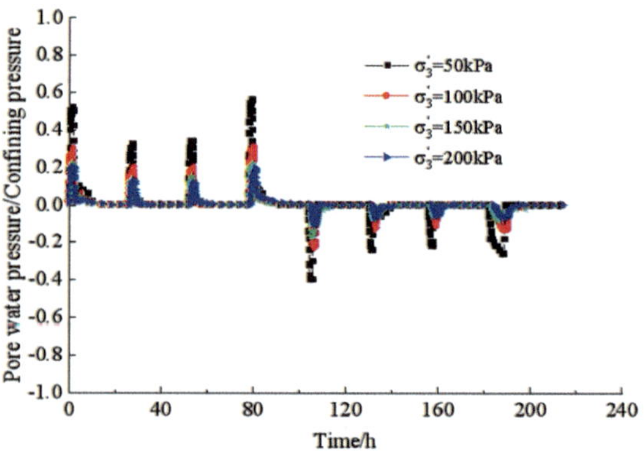

Table 5.4 Normalized pore water pressure and volumetric strain change caused by inside and outside boundary isothermal multi-stage heating–cooling process

Confining pressure (kPa)	Determination of physical quantities	Heating (°C)				Cooling (°C)			
		40	50	60	75	60	50	40	25
50	$(u/\sigma_3')_{peak}$	0.48	0.32	0.34	0.54	− 0.4	− 0.24	− 0.22	− 0.26
	$\epsilon_v/\%$	0.51	1.01	1.69	2.71	2.21	1.74	1.25	0.76
100	$(u/\sigma_3')_{peak}$	0.30	0.2	0.2	0.31	− 0.22	− 0.12	− 0.11	− 0.13
	$\epsilon_v/\%$	0.54	0.97	1.65	2.75	2.18	1.70	1.20	0.79
150	$(u/\sigma_3')_{peak}$	0.21	0.13	0.15	0.22	− 0.16	− 0.07	− 0.06	− 0.08
	$\epsilon_v/\%$	0.53	0.98	1.54	2.51	2.31	2.08	1.72	1.33
200	$(u/\sigma_3')_{peak}$	0.19	0.125	0.115	0.17	− 0.11	− 0.065	− 0.055	− 0.065
	$\epsilon_v/\%$	0.61	1.16	1.74	2.60	2.23	1.94	1.58	1.04

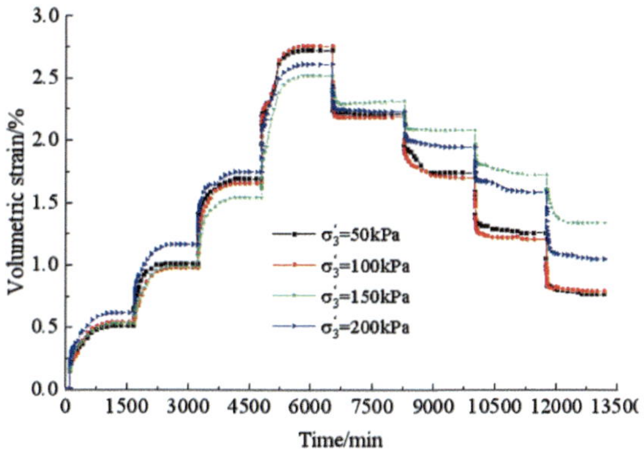

Fig. 5.24 Volumetric strain change during inside and outside boundary isothermal multi-stage heating–cooling process

temperature rise and fall but cannot be used to measure the change in strain of the sample body.

(3) Influence of Confining Pressure and Temperature on Pore Pressure and Volumetric Strain

Based on the data in the table, the following observations can be made regarding changes in pore pressure: During step-by-step heating with constant confining pressure, the normalized pore water pressure achieved for the same temperature increment is consistent. For example, heating from 40 °C to 50 °C and from 50 °C to 60 °C under the same confining pressure results in similar pore pressures (150 kPa yielding 0.13 and 0.15, respectively; 100 kPa yielding 0.28 and 0.31, respectively). Larger temperature increments result in higher pore pressures, and for the same temperature increment, the pore pressure is not significantly affected by the specific temperature values. During cooling, the pore pressures achieved for the same temperature increment are similar under the same confining pressure, such as cooling from 60 °C to

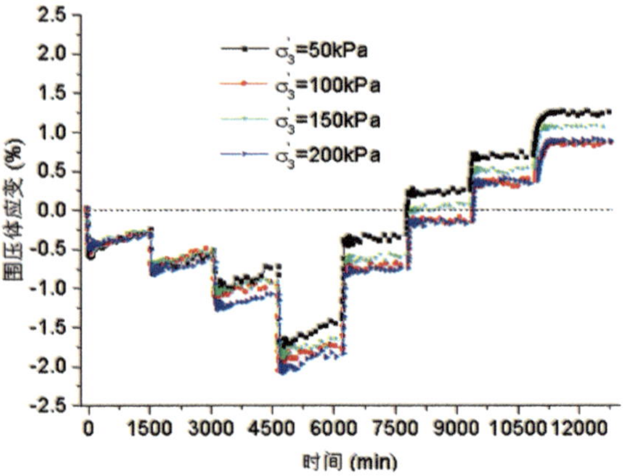

Fig. 5.25 Confining pressure volumetric strain change during inside and outside boundary isothermal multi-stage heating–cooling process

50 °C and from 50 °C to 40 °C (50 kPa, − 0.24 and − 0.22, respectively). However, for cooling from 75 °C to 60 °C and from 40 °C to 25 °C, although the temperature increments are the same, the resulting pore pressures differ significantly (200 kPa, − 0.105 and − 0.065, respectively). Larger cooling increments result in higher pore pressures, and for the same temperature increment, higher temperatures tend to result in larger pore pressures.

Regarding changes in volumetric strain: During step-by-step heating with constant confining pressure, for the same temperature increment, higher temperatures result in larger volumetric strains. For example, at 100 kPa confining pressure, heating from 40 °C to 50 °C and from 50 °C to 60 °C results in volumetric strains of 0.43% and 0.68%, respectively; heating from 25 °C to 40 °C and from 60 °C to 75 °C results in strains of 0.54% and 1.1%, respectively. Larger temperature increments lead to larger volumetric strains. During step-by-step cooling with constant confining pressure, the trend is opposite to that of heating. For the same temperature increment, higher temperatures tend to result in smaller volumetric strains. For instance, at 150 kPa confining pressure, cooling from 60 °C to 50 °C and from 50 °C to 40 °C results in strains of 0.23% and 0.36%, respectively; cooling from 75 °C to 60 °C and from 40 °C to 25 °C results in strains of 0.20% and 0.39%, respectively. At 100 kPa, cooling from 75 °C to 60 °C and from 40 °C to 25 °C results in strains of 0.57% and 0.41%, respectively, indicating that higher temperatures yield larger strains. Overall, for the same temperature increment during cooling, higher temperatures tend to result in smaller strains, and larger cooling increments result in larger strains.

From Fig. 5.26, it can be observed that for step-by-step heating and cooling to the same temperature, a higher confining pressure results in a lower maximum normalized pore water pressure achieved. In Fig. 5.26a, for the same temperature increment, the changes are generally similar (e.g., heating from 50 °C to 60 °C and from 40 °C to 75 °C with increments of 10 °C and 15 °C respectively show similar trends).In Fig. 5.26b, during cooling, larger temperature increments result in higher pore pressures, and with increasing confining pressure, the patterns of change for different cooling increments become more similar.

Fig. 5.26 Relationship between pore water pressure and confining pressure during isothermal multi-stage heating–cooling process

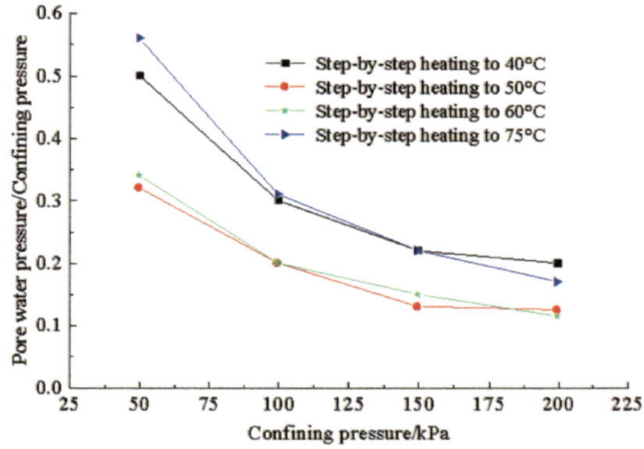

(a) Isothermal multistage heating process

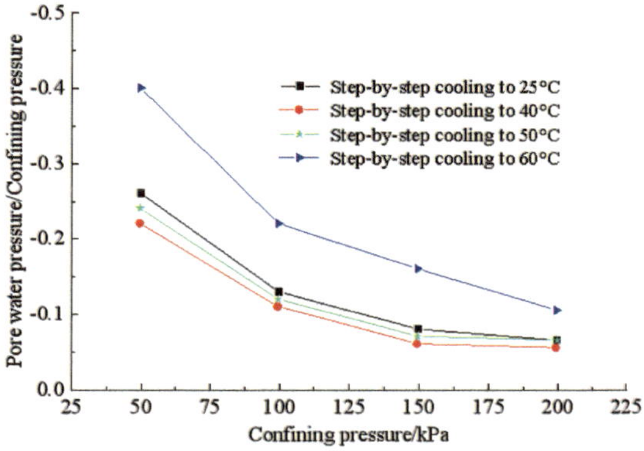

(b) Isothermal multistage cooling process

Figure 5.27 shows the relationship between volumetric strain and temperature during isothermal multi-stage heating and cooling under different confining pressures. After initial consolidation is completed, the volumetric strain is defined as 0. Subsequently, during step-by-step heating, the volumetric strain continuously increases. It can be observed that the volumetric strain under different confining pressures is similar and exhibits a linear growth. During step-by-step cooling, the volumetric strain decreases. The changes at 50 and 100 kPa confining pressures are quite similar, while at 150 and 200 kPa confining pressures, the volumetric strain shows a larger difference compared to 50 kPa and changes more slowly.

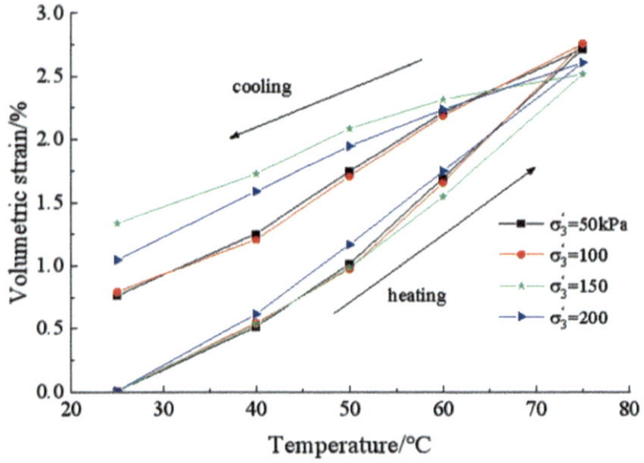

Fig. 5.27 Volumetric strain change during isothermal multi-stage heating–cooling process

5.4 Thermal Consolidation Tests on Hollow Cylindrical Samples Under Non-isothermal Boundary Conditions

5.4.1 Test Plan

Based on the aforementioned experimental setup, the design of thermal consolidation tests under non-isothermal boundary conditions focuses on the evolution of average pore water pressure and consolidated volumetric strain inside the sample over time, under different radial temperature gradients (controlled by the inner and outer boundaries). The radial pressures applied in the experiments include four levels: 50, 100, 150, and 200 kPa. During testing, the temperature range for the inner and outer boundaries is 25–75 °C. Initially, the sample is consolidated at the environmental temperature of 25 °C under a given radial pressure. Then, temperature loading is gradually applied through the inner and outer boundaries of the sample, following three different heating paths (see Table 5.5; Fig. 5.28). During the boundary heating process, conditions are undrained; once the temperature reaches a stable set point, the sample is allowed to consolidate with drainage.

5.4.2 Heating Process of the Inner and Outer Boundaries of the Sample

Figure 5.29 shows the temperature changes of the inner and outer boundaries of the sample under a given consolidation pressure after initial consolidation ($T_0 = 25$ °C) for three different heating paths. It should be noted that due to the relatively high heating power of the heating rod at the center of the sample, the temperature rise at the inner boundary is steeper. The process of pore water pressure dissipation after initial consolidation is shown in Fig. 5.30a. For example, considering Heating Path 1 (with

Table 5.5 Schemes of heating specimens

Heating paths	Consolidation pressure (σ_3/kPa)	Initial temperature T_0/%	Heating sequence	Inner boundary temperature T_1/°C	Outer boundary temperature T_2/°C	Temperature gradient (°C cm^{-1})
1	50, 100, 150, 200	25	1	50	40	2.50
			2	75	60	3.75
			3	75	75	0
2	50, 100, 150, 200	25	1	75	60	3.75
			2	75	75	0
3	50, 100, 150	25	1	50	40	2.50
			2	50	50	0
			3	75	60	3.75
			4	75	75	0
4	100	25	1	50	40	2.50
			2	50	50	0
			3	75	60	3.75
			4	75	75	0
			5	25	25	0
5	100	25	1	40	50	2.50
			2	50	50	0
			3	60	75	3.75
			4	75	75	0
			5	50	50	0
			6	25	25	0

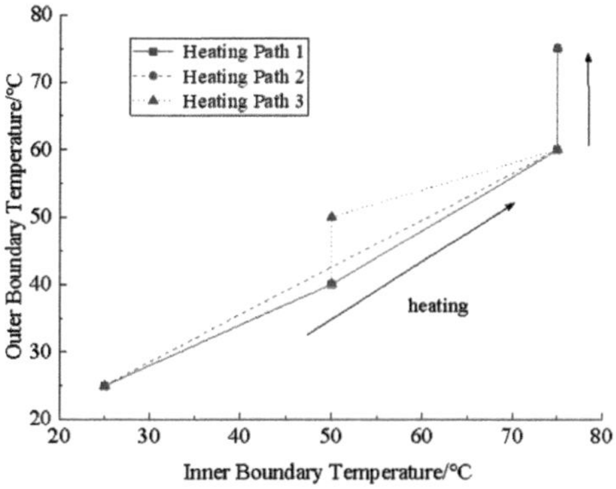

Fig. 5.28 Three heating paths for consolidation tests

three heating levels) as illustrated in Fig. 5.29a and b (with confining pressures σ_3 = 50 kPa and 150 kPa respectively), the following can be observed: At the first heating level, when the inner boundary temperature is increased to $T_1 = 50$ °C (approximately 30 min), the temperature of the outer boundary gradually rises and stabilizes at $T_2 = 40$ °C after about 200 min, resulting in an inner-to-outer radial temperature gradient of $\Delta T = 10$ °C, which corresponds to 2.50 °C/cm. Once consolidation is completed under this temperature gradient, the second heating level is applied, raising the inner boundary temperature to $T_1 = 75$ °C (approximately 20 min), while the outer boundary temperature gradually rises to $T_2 = 60$ °C, resulting in an inner-to-outer boundary temperature difference of $\Delta T = 15$ °C and a temperature gradient of 3.75 °C/cm, and further consolidation occurs under this gradient. Subsequently, the third heating level is applied, raising the outer boundary temperature to $T_2 = 75$ °C, and the sample reaches a uniform temperature state again. Comparison of Fig. 5.29a and b indicates that the temperature change processes are similar under different consolidation pressures (e.g., $\sigma_3 = 50$ kPa and $\sigma_3 = 150$ kPa), which is consistent with conclusions found in existing literature [4].

For Heating Paths 2 and 3 (with 2 and 4 heating levels respectively, as shown in Fig. 5.28), the processes differ from Heating Path 1, but the temperature increase is similar. Ultimately, all samples are heated to an isothermal state of $T_1 = T_2 = 75$ °C.

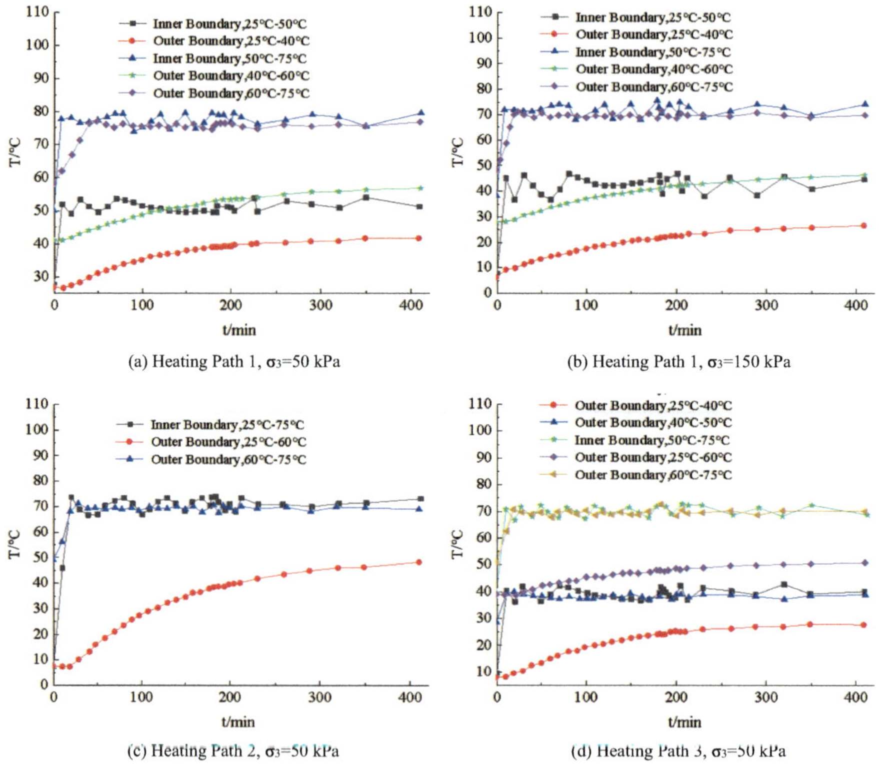

Fig. 5.29 Variation of temperatures of inner and outer boundaries of specimens

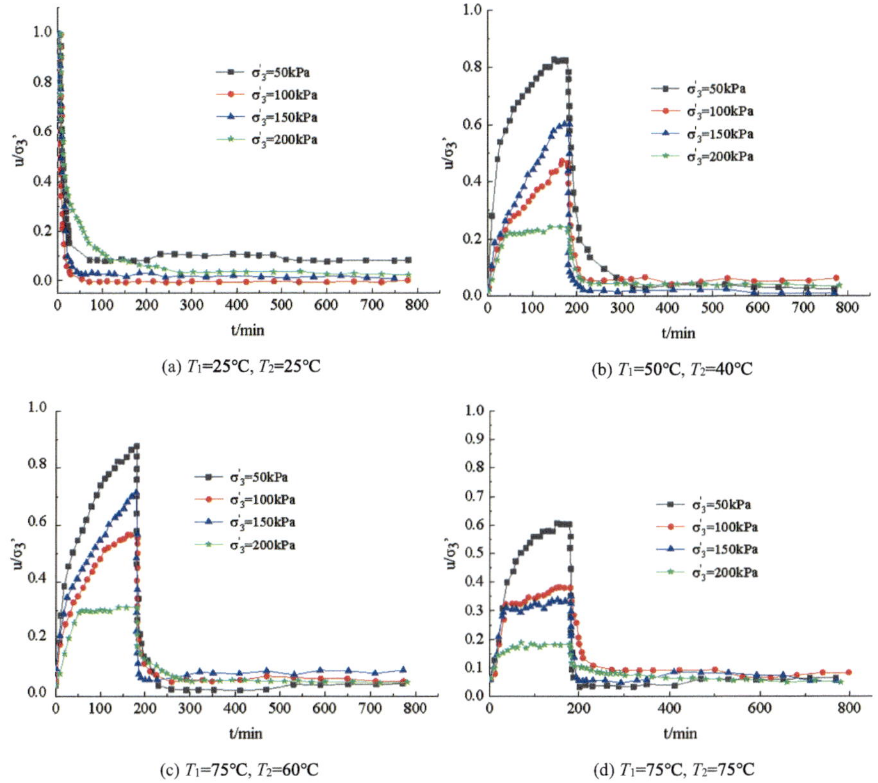

Fig. 5.30 Evolution of pore pressures with time (Heating Path 1)

5.4.3 Changes in Pore Water Pressure

Figures 5.30, 5.31, and 5.32 show the growth and dissipation of pore water pressure inside the sample under three different heating paths, with heating applied to the inner and outer boundaries of the sample. From Fig. 5.30a, it can be observed that under isothermal conditions ($T_0 = 25$ °C) and radial pressures (including $\sigma_3 = 50$, 100, 150, 200 kPa), the pore water pressure dissipates quickly. After 200 min, the pore water pressure becomes very small (approaching $u/\sigma_3 = 0$) or stabilizes.

The experiments show that even after significant temperature effects and consolidation have occurred, further increasing the temperature range still leads to a notable rise in pore water pressure. Additionally, the evolution of pore water pressure under the three heating paths indicates that as the consolidation pressure increases, the normalized pore water pressure u/σ_3 decreases, and the rate of increase in pore water pressure tends to slow down, while the actual pore water pressure generally increases. For example, for Heating Path 2 (Fig. 5.31b), where the temperature increment for the second level is only 15 °C (see Table 5.5), the pore water pressure can still reach

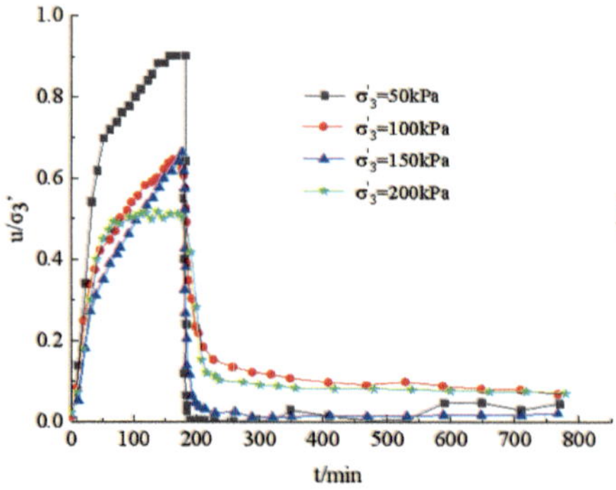

(a) T_1=75°C, T_2=60°C

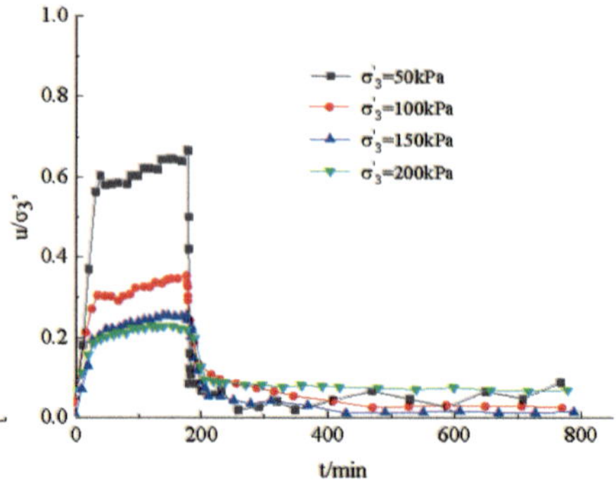

(b) T_1=75°C, T_2=75°C

Fig. 5.31 Evolution of pore pressures with time (Heating Path 2)

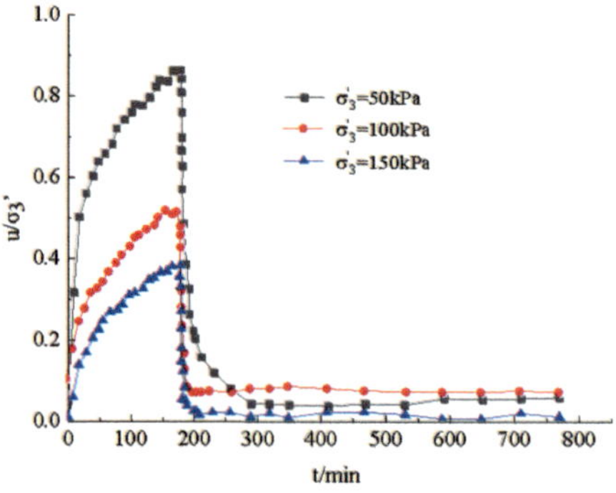

(a) T_1=50°C, T_2=40°C

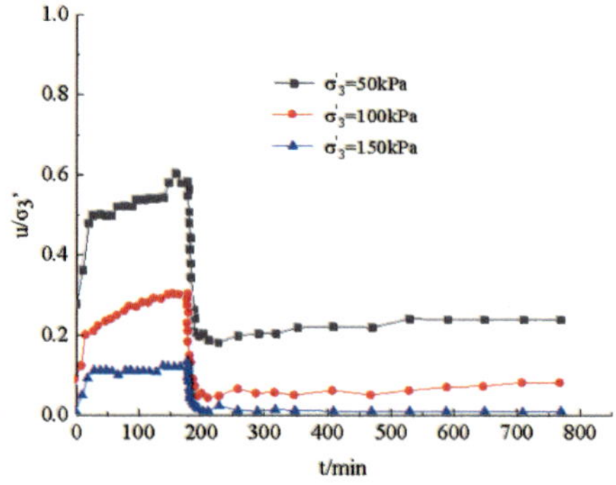

(b) T_1=50°C, T_2=50°C

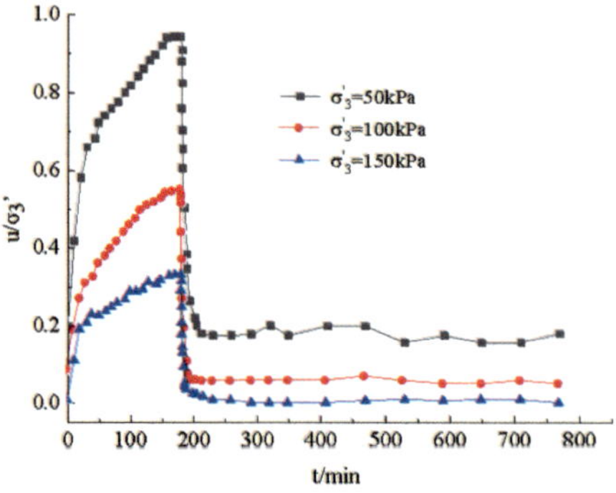

(c) T_1=75°C, T_2=60°C

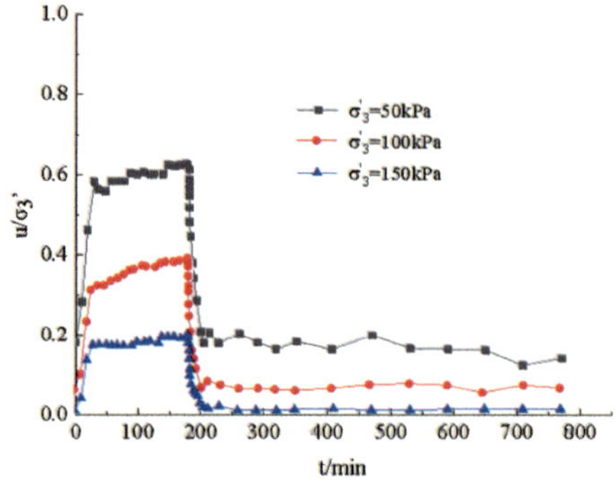

(d) T_1=75°C, T_2=75°C

Fig. 5.32 Evolution of pore pressures with time (Heating Path 3)

$u/\sigma_3 = 0.66, 0.34, 0.25, 0.23$ for $\sigma_3 = 50, 100, 150, 200$ kPa, respectively (with actual pore water pressures of u = 33.0, 34.0, 37.5, 45.6 kPa).

Figure 5.33 illustrates the evolution of pore pressure and temperature under a confining pressure of 100 kPa when the internal heat source is greater than the external heat source. After initial consolidation, the internal heat source temperature is raised to 50 °C, and the external heat source temperature is raised to 40 °C (actual 42.8 °C). The normalized pore pressure reaches 0.41. Following consolidation and drainage, the pore pressure dissipates, and the temperature remains relatively stable with minimal fluctuations. Subsequently, with the internal heat source temperature unchanged, the external heat source temperature is increased from 40 to 50 °C. Due to the small temperature increment, equilibrium is reached in 60 min, with a low normalized pore pressure of 0.1. After consolidation and drainage, further heating raises the internal heat source to 75 °C and the external heat source to 60 °C (actual 63 °C). The normalized pore pressure reaches 0.35, with a relatively stable temperature difference. After consolidation and drainage, the internal heat source temperature is held constant while the external heat source is increased to 75 °C. The normalized pore pressure reaches 0.19, with a heating duration of 80 min. After consolidation and drainage, cooling is performed without drainage, with both internal and external heat sources reduced directly to 25 °C. The cooling process for both sources is nearly identical, taking approximately 440 min, with the normalized pore pressure reaching − 0.48. After drainage and consolidation, the pore pressure dissipates to zero. The drainage and consolidation time for each step is 1440 min to ensure complete pore pressure dissipation.

Figure 5.34 shows the evolution of pore pressure and temperature under a confining pressure of 100 kPa when the external heat source is greater than the internal heat source. After initial consolidation, the external heat source is raised to 50 °C and the internal heat source to 40 °C (actual 48.5 °C). The temperatures are very close and show significant fluctuations, with the normalized pore pressure reaching 0.52. After consolidation and drainage, with the external heat source temperature held constant, the internal heat source is raised to 50 °C, causing minimal changes in

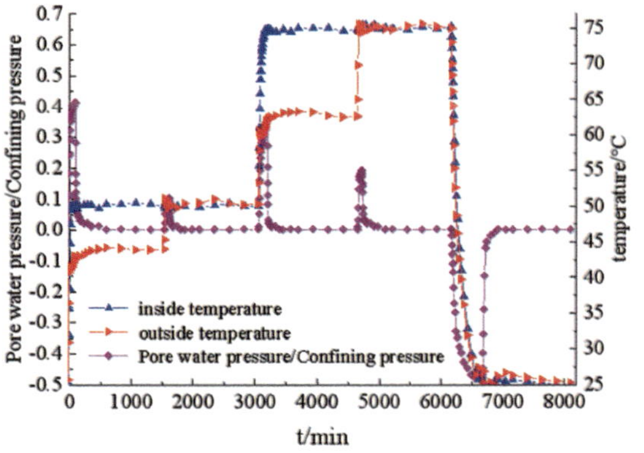

Fig. 5.33 Pore water pressure and temperature change when the outside temperature is greater than the inside at 100 kPa

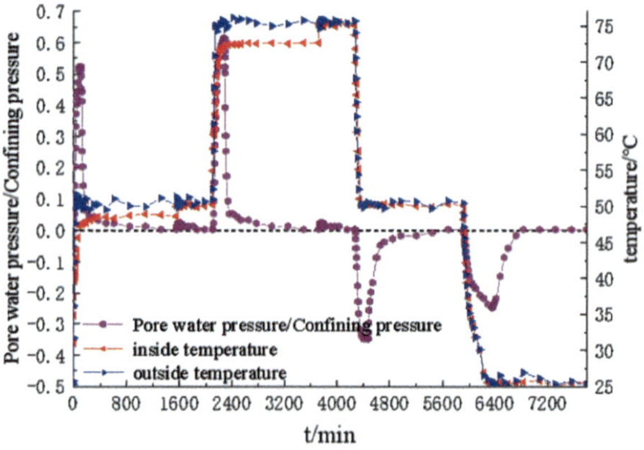

Fig. 5.34 Pore water pressure and temperature change when the inside temperature is greater than the outside at 100 kPa

pore pressure, which drops to 0.02. The heating and drainage times are correspondingly reduced. Once the pore pressure is fully dissipated, the external heat source is increased to 75 °C, and the internal heat source to 60 °C (actual 72 °C), with the normalized pore pressure reaching 0.61. After consolidation and drainage, with the external heat source temperature unchanged, the internal heat source is further increased to 75 °C, resulting in minimal pore pressure change, stabilizing at 0.02. Subsequent cooling is performed in two stages: first, both sources are cooled to 50 °C, resulting in a normalized pore pressure of − 0.35. After consolidation and drainage, the temperature is further reduced simultaneously to 25 °C, with the normalized pore pressure reaching − 0.25. Complete drainage and consolidation ensure the pore pressure fully dissipates.

5.4.4 Changes in Volumetric Strain

For the soil samples prepared in this study, at confining pressures of $\sigma_3 = 50$, 100, 150, and 200 kPa, the consolidated volumetric strains under isothermal conditions ($T_0 = 25$ °C) are $\varepsilon_v = 2.92\%$, 4.05%, 6.12%, and 7.27%, respectively (Fig. 5.35). The dissipation process of the pore water pressure is shown in Fig. 5.30a.

Then, applying boundary temperature loads results in very rapid pore water pressure dissipation during the drainage consolidation process after each temperature increase. Typically, pore water pressure stabilizes within 60 min. At this point, the sample experiences significant thermal consolidation volumetric strain due to the dissipation of pore water pressure (Fig. 5.36). In Fig. 5.36, Heating Paths 1, 2, and 3 represent the drainage consolidation process after three, two, and four instances of non-drained boundary heating, respectively, with levels 1, 2, 3, etc., indicating the sequence of heating levels within a given path. The figure shows that after the boundary temperatures reach the set values, the drainage consolidation process causes

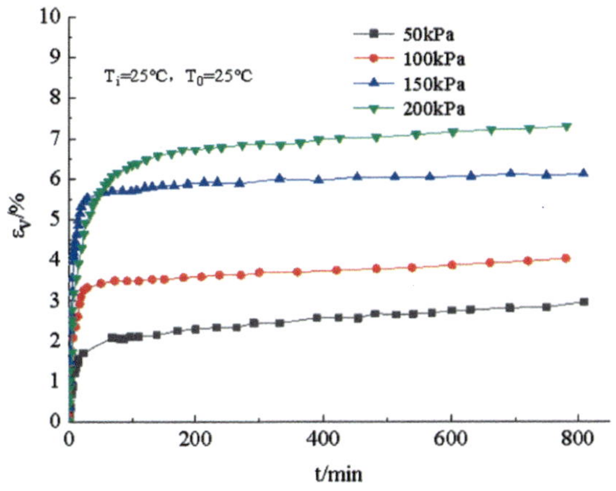

Fig. 5.35 Consolidation of specimens under isothermal condition

the pore water pressure to dissipate, activating the bound water around soil particles, resulting in new volumetric strain. As the pore water pressure dissipates, the volumetric strain of the sample continues to increase and eventually stabilizes.

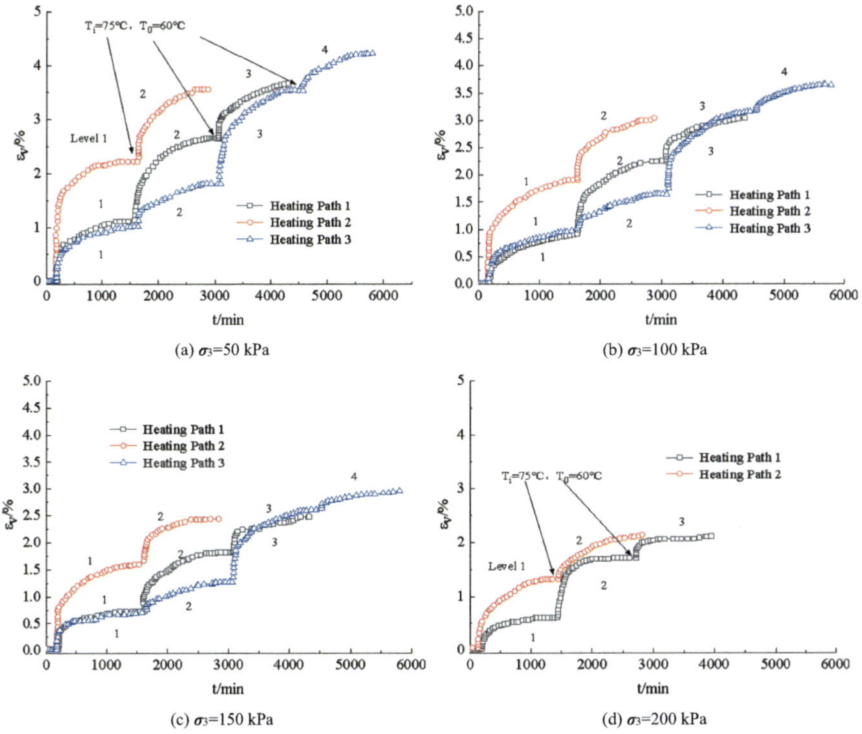

Fig. 5.36 Evolution of volumetric strains with time under various heating paths

With an increase in heating levels (according to heating paths 3, 1, 2, corresponding to 4, 3, 2 levels, respectively, see Table 5.5), the final consolidation volumetric strain increases even at the same higher temperature (e.g., $T_1 = T_2 = 75$ °C, isothermal state). Comparing Fig. 5.36a–d, it is evident that with increasing confining pressure, the consolidation volumetric strain under the same heating path conditions decreases. For example, at confining pressures of $\sigma_3 = 50$, 100, and 150 kPa, under Heating Path 1 (Fig. 5.36), the volumetric strain caused by the first level of heating is $\varepsilon_v = 1.12\%$, 0.92%, and 0.72%, respectively; the strain increments caused by the second level of heating are $\varepsilon_v = 1.54\%$, 1.31%, and 1.09%, resulting in total strains of $\varepsilon_v = 2.66\%$, 2.23%, and 1.81%, respectively. In practice, higher confining pressures result in larger volumes of water expelled during the initial isothermal consolidation stage ($T_0 = 25$ °C), leading to smaller pore water pressures and volumetric strains induced by subsequent temperature loads.

Figure 5.36 shows that although the final temperature values are the same for all three heating paths (see Table 5.5, all at $T_1 = T_2 = 75$ °C), the final volumetric strain differs. It appears that the more temperature levels set during heating, the greater the total consolidation strain. For instance, when $\sigma_3 = 50$ kPa (Fig. 5.36a), the final strains for the three heating paths with increasing levels (i.e., Path 2 → 1 → 3) are $\varepsilon_v = 3.56\%$, 3.67%, and 4.25%, respectively. This reflects the impact of the temperature load path on the sample's thermal consolidation effect. Even in intermediate heating processes, when the temperature gradient is the same (e.g., inner boundary temperature at $T_1 = 75$ °C and outer boundary temperature at $T_2 = 60$ °C, as marked in Fig. 5.36), the volumetric strain increases with more heating levels (i.e., Path 2 → 1 → 3). For example, at $\sigma_3 = 50$ kPa, the strains for these three heating paths are $\varepsilon_v = 2.22\%$, 2.66%, and 3.53%, respectively. In reality, due to the temperature load, the increase in pore water pressure generally does not exceed the consolidation pressure the sample is subjected to, so the increase in pore water pressure is not linear. That is, with the same temperature increment, the more temperature levels set, the greater the cumulative increase in pore water pressure, resulting in more noticeable volumetric strain due to the dissipation of pore water pressure. This effect is more pronounced at lower consolidation pressures and tends to converge at higher pressures (see Fig. 5.36d).

Figure 5.37 shows the changes in volumetric strain during non-isothermal heating of the inner and outer boundaries:

(a) Inner temperature greater than outer temperature: Due to a noticeable temperature difference, the changes during the four heating stages are quite pronounced. The cooling process involves simultaneous cooling of both inner and outer boundaries to 25 °C.
(b) Outer temperature greater than inner temperature: Because the actual temperature difference in non-isothermal control is relatively small, there is minimal change during the two heating stages (inner temperatures at 48.5 and 50 °C, 72 and 75 °C). In contrast, the other two stages with larger actual temperature increments show relatively larger changes in volumetric strain. The cooling process

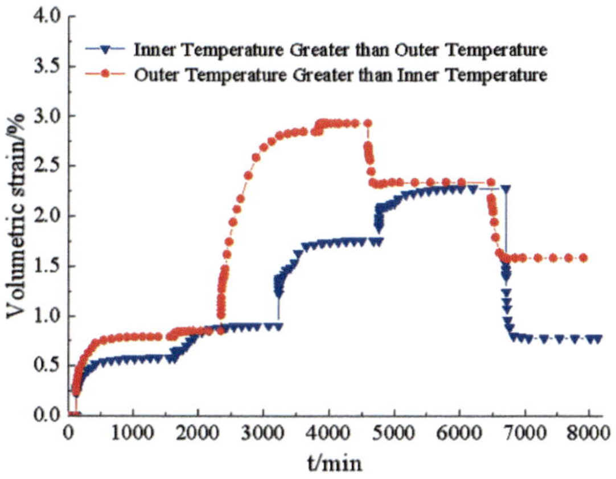

Fig. 5.37 Volumetric strain change during non-isothermal multi-stage heating–cooling process

is divided into two stages, with both boundaries cooling simultaneously first to 50 °C and then to 25 °C.

5.4.5 Comparative Analysis

(1) Comparison of Pore Pressure Evolution

Figure 5.38 presents the normalized pore pressure changes under different heating and cooling conditions with a confining pressure of 100 kPa. Overall, it can be observed that the greater the temperature increment, the higher the achieved pore pressure. For example, with single-stage heating (inner and outer temperatures both 50 °C), the pore pressure reached 0.53. When the outer temperature was higher than the inner temperature (outer 50 °C, inner 48.5 °C), the pore pressure was 0.52. For the case where the inner temperature was higher than the outer temperature (outer 42.8 °C, inner 50 °C), the pore pressure was 0.41. Under equal temperature (40 °C), the pore pressure was 0.3. This data clearly shows that, for the same temperature increment, higher boundary temperatures result in greater pore pressures. For single-stage cooling to 75 °C, the pore pressure reached – 0.50. In the case where the inner temperature was higher than the outer temperature during cooling (75 °C inner, 25 °C outer), the pore pressure was – 0.48. Single-stage cooling to 50 °C resulted in a pore pressure of – 0.19, while cooling with the outer temperature greater than the inner temperature (50 °C outer, 25 °C inner) resulted in a pore pressure of – 0.25. Cooling to 40 °C yielded a pore pressure of – 0.13, and cooling with equal temperatures (40 °C) resulted in a pore pressure of – 0.12. It can be seen that the pore pressure achieved with sequential and single-stage cooling processes is roughly similar and is not significantly affected by the staged relationships.

(2) Comparison of Volumetric Strain Evolution

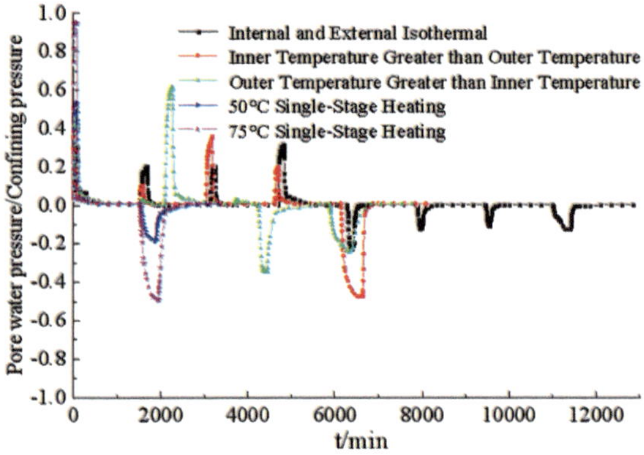

Fig. 5.38 Comparisons of pore water pressure change of different paths at 100 kPa

Figure 5.39 presents a comparison between the volumetric strain changes from single-stage and multi-stage heating and cooling. Overall, it is observed that the volumetric strain produced by multi-stage is greater than that produced by single-stage heating at the same temperature. This phenomenon can be attributed to the longer duration of thermal consolidation during multi-stage heating, which activates more bound water within the sample and allows for more thorough drainage due to the multiple drainage stages. The total strain resulting from single-stage heating and cooling at 75 °C differs from that produced by multi-stage (up to 75 °C) heating and cooling and is independent of the confining pressure.

Figure 5.40 shows the comparison of volumetric strain under a confining pressure of 100 kPa, where the temperature is increased from the ambient temperature of 25–75 °C and then back to 25 °C, for different heating paths. When the inner and outer heat sources are raised to 50 °C through different paths, the resulting volumetric strains are as follows: staged isothermal heating is 0.97%, staged inner temperature greater than outer temperature is 0.90%, staged outer temperature greater than inner temperature is 0.85%, and single-stage heating to 50 °C is 0.66%. When heating the inner and outer heat sources to 75 °C through different paths, the corresponding volumetric strains are: staged isothermal heating is 2.75%, staged outer temperature greater than inner temperature is 2.62%, staged inner temperature greater than outer temperature is 2.28%, and single-stage heating to 75 °C is 0.67%. This indicates that different heating paths to the same temperature produce different volumetric strains, with multi-stage heating resulting in greater strain compared to single-stage heating, and staged isothermal heating yielding greater strain compared to non-isothermal heating.

In the cooling process where both inner and outer heat sources are cooled isothermally to 50 °C, the results are as follows: cooling from 75 to 50 °C produces a volumetric strain of 0.59%, while cooling from 75 to 50 °C and then to 50 °C results in a strain of 1.05%. For cooling to 25 °C, the results are: cooling from 75 to 25 °C produces a strain of 1.48%, while staged inner temperature greater than outer temperature results in 1.50%; cooling from 75 to 50 °C and then to 25 °C yields a strain of 1.35%, and staged cooling from 75 to 50 °C and then to 40 °C and finally to 25 °C

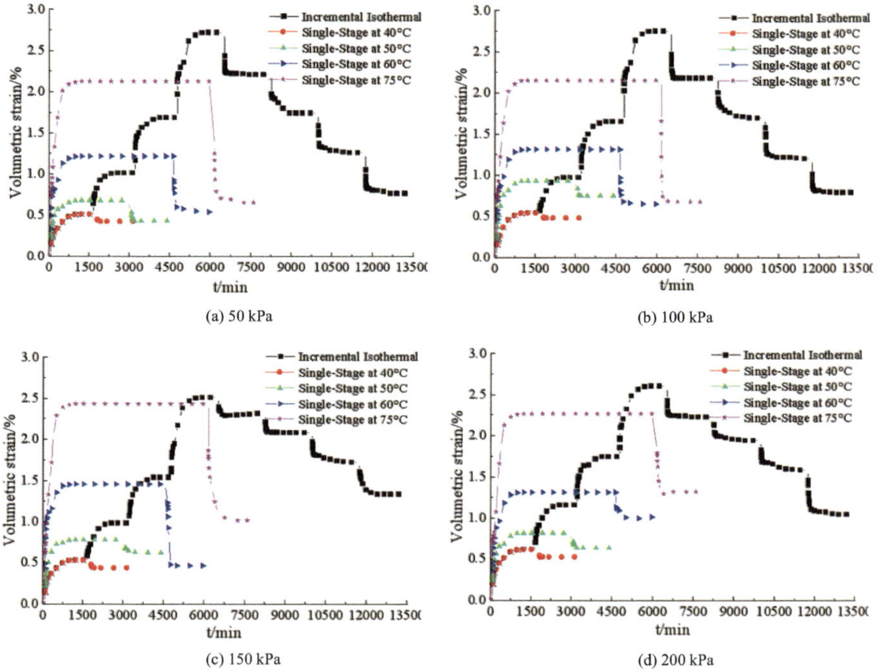

(a) 50 kPa (b) 100 kPa (c) 150 kPa (d) 200 kPa

Fig. 5.39 Comparisons of volumetric strain change of isothermal single-stage and multi-stage heating–cooling process

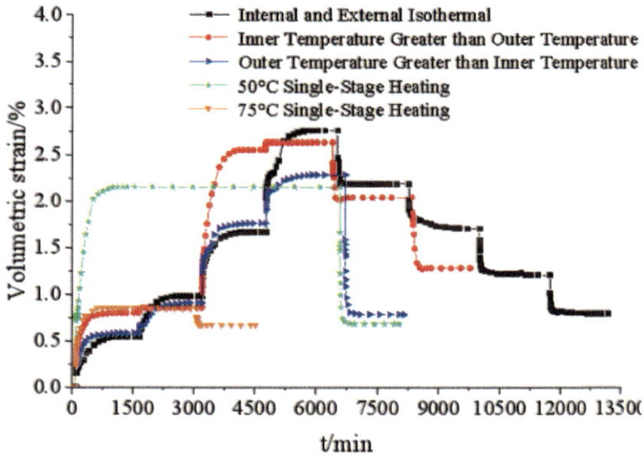

Fig. 5.40 Comparison of volumetric strain change of various kinds of paths

results in a strain of 1.96%. This indicates a trend of increasing volumetric strain with staged cooling. However, when cooling to 25 °C, staged cooling results in a smaller strain compared to single-stage cooling, possibly due to the natural cooling rate which is difficult to control.

After the heating and cooling processes, the final volumetric strains are as follows: staged isothermal heating and cooling results in 0.79%, staged outer temperature greater than inner temperature results in 1.27%, staged inner temperature greater than outer temperature results in 0.78%, and single-stage heating and cooling to 75 °C results in 0.67%. This indicates that the final volumetric strain varies with different heating and cooling paths, and there is no discernible pattern.

5.5 Summary

Using the self-developed thermal consolidation testing device, a comprehensive study on the thermal consolidation characteristics of saturated clay under different inner and outer boundary temperature conditions for hollow cylindrical specimens was conducted. The temperature, pore water pressure, and consolidation strain variations throughout the test process were analyzed, leading to the following conclusions:

(1) Initial Isothermal (25 °C) Consolidation: The pore water pressure dissipation curves and consolidation strain change curves over time are quite similar across different confining pressures. Additionally, higher confining pressure results in a slower dissipation rate of pore pressure, greater volume of drainage, larger consolidation strain, and a longer time to reach stability. Temperature fluctuations are independent of the confining pressure.
(2) Isothermal Single-Stage Heating and Cooling Process: During the drainage process, pore water pressure induced by temperature changes rapidly dissipates and stabilizes more quickly compared to the initial consolidation process. For the same temperature path, the pore water pressure variation is consistent across different temperature increments. The maximum normalized pore water pressure achieved decreases with increasing confining pressure; conversely, for the same confining pressure, the maximum normalized pore water pressure increases with greater temperature increments until reaching a potential maximum value of 1. Under the same temperature load, the rate of increase in pore water pressure slows as the confining pressure increases. Regarding strain changes, after heating and drainage, the consolidation strain of saturated soil rapidly increases and quickly stabilizes. Following cooling and drainage, due to the development of negative pore pressure and water absorption phenomena, the strain decreases from the previous value but overall increases. For the same temperature path, consolidation strain is independent of confining pressure; for the same confining pressure, greater temperature increments result in larger consolidation strains.
(3) Isothermal Multi-Stage Heating and Cooling Process: The patterns observed in pore water pressure and consolidation strain changes are similar to those in single-stage heating and cooling processes. Within a certain temperature range (25–75 °C), as the confining pressure remains constant, the pore pressure increases with larger temperature increments during heating. For the same

increment, the achieved pore pressure remains similar, regardless of temperature. During cooling, larger temperature decrements result in higher pore pressures, with greater pore pressures at higher temperatures for the same decrement. Regarding strain changes, for a given confining pressure, the consolidation strain increases with higher temperature increments during heating, while it tends to decrease with higher temperature increments during cooling.

(4) Non-Isothermal Boundary Conditions: Different heating levels and sequences result in significant variations in the maximum stabilized pore water pressure under the same temperature load. Additionally, after a significant temperature effect and consolidation, further increases in temperature increments still lead to substantial rises in pore water pressure.

(5) Effect of Heating Levels on Consolidation Strain: As the number of heating levels increases, the consolidation strain of the specimen increases. Conversely, for the same heating path, higher confining pressures result in smaller consolidation strains. This is because, under higher confining pressures, the consolidation strain generated during the initial isothermal consolidation phase is greater, while the subsequent temperature-induced pore water pressure and strain effects are smaller.

(6) Effect of Heating Pathways on Strain: Even with the same final temperature, different heating paths result in varying consolidation strains. Generally, more temperature levels set during heating lead to greater total consolidation strain. In reality, with the same temperature increment conditions, more temperature levels result in greater cumulative pore water pressure increases, making the strain due to pore water pressure dissipation more pronounced. This highlights the impact of temperature paths on the thermal consolidation effects of the specimen.

References

1. Bai B, Li T (2009) Solutions for cylindrical cavity in saturated thermoporoelastic medium. Acta Mech Solida Sin 22(1):85–94
2. Bai B (2013) Thermal response of saturated porous spherical body containing a cavity under several boundary conditions. J Therm Stresses 36(11):1217–1232
3. Demars KR, Charles RD (1981) Soil volume change induced by temperature cycling. Can Geotech J 19(2):189–194
4. Bai B, Guo LJ, Han S (2014) Pore pressure and consolidation of saturated silty clay induced by progressively heating/cooling. Mech Mater 75:84–94

Chapter 6
Thermal Consolidation of Saturated Silty Clay Considering Overconsolidation Effect with Different Heating–Cooling Paths

6.1 Experiment Overview

(1) Experimental Preparation

Experiments were conducted using a self-developed thermal consolidation device suitable for hollow cylindrical specimens. The tested soil was silty clay commonly found in the northern regions, obtained through core drilling in the northwest part of Beijing. After drying, crushing, and sieving through a 0.5 mm screen, the remolded soil samples were prepared according to the standard with a moisture content of 15% and dry density of 1.66 g/cm^3 (Fig. 6.1). The basic physical properties of the samples are shown in Table 6.1.

(2) Experiment Plan 1—effects of temperature paths on soils with different overconsolidation ratios

The overconsolidation ratios of the specimens used in the experiment are 1, 4, and 8. Prior to conducting the experiment, calibration of the instruments is necessary, followed by the installation of the specimens. After preparation, proceed with initial consolidation. Consolidation pressures of 50, 200, and 400 kPa will be applied at a reference temperature of 25 °C for 24 h to ensure complete dissipation of pore water pressures. Subsequently, for specimens subjected to consolidation pressures of 200 and 400 kPa, the confining pressure will be reduced to 50 kPa under undrained conditions. After allowing pore pressure to stabilize (approximately 2 h), drainage will then be initiated until pore pressures dissipate completely, over a period of 24 h. This process will yield soil samples with overconsolidation ratios of 4 and 8. After completing the initial consolidation, the experiment follows the heating and cooling paths outlined in Table 6.2. The confining pressure remains constant at 50 kPa throughout. The procedure begins with undrained heating to the specified temperature, maintained for 2 h. Subsequently, the temperature is held constant while conducting consolidation drainage until pore pressures dissipate over 24 h; Following

G. Yang and B. Bai, *Experimental and Theoretical Study on Thermal Consolidation Behavior of Saturated Clay*, https://doi.org/10.1007/978-981-96-6909-7_6

Fig. 6.1 Test sample

Table 6.1 Basic physical parameters of soil sample

Initial porosity	Saturation degree (%)	Specific gravity	Dry density (g/cm^3)	Liquid limit (%)	Plastic limit (%)	Plasticity index (%)
0.64	98.0	2.72	1.66	29.95	16.05	13.9

this, the specimen undergoes undrained natural cooling to the specified temperature over 8 h. Then, consolidation drainage is applied again until pore pressures dissipate completely, taking another 24 h. In Table 6.2, Path 1 and Path 2 represent single-stage heating and cooling processes, while Path 3 involves a multi-stage heating and cooling process. Each stage follows the aforementioned procedures accordingly.

(3) Experiment 2—effect of Consolidation Pressure on Different Overconsolidation Ratios of Soil Samples.

Under 25 °C conditions, the procedure involves gradually increasing the confining pressure to a specific value (e.g., 100, 150, 200, 300 kPa) to achieve initial consolidation. Subsequently, the confining pressure is reduced to 50, 100, 150, or 200 kPa for the process of wetting consolidation until stability is reached. This results in the formation of overconsolidated soil samples with different Overconsolidation Ratios (OCR) such as 1, 2, 3, 4, etc. The sample is heated to 75 °C under undrained conditions, then allowed to consolidate under drainage to dissipate pore water pressures. Finally, the sample is cooled to 25 °C under undrained conditions and allowed to absorb water, while monitoring the evolution of pore water pressures and consolidation strain.

Table 6.2 Test scheme

Heating–cooling path	OCR	Internal heat source (°C)	External heat source (°C)
1	1, 4, 8	25 → 50 → 25	25 → 50 → 25
2	1, 4, 8	25 → 75 → 25	25 → 75 → 25
3	1, 4, 8	25 → 40 → 50 → 60 → 75 → 60 → 50 → 40 → 25	25 → 40 → 50 → 60 → 75 → 60 → 50 → 40 → 25

6.2 Analysis of Experimental Results—Method 1

6.2.1 *Initial Consolidation Process*

In the experiment, the volumetric strain refers to the back-pressure volumetric strain, which is the drainage volumetric strain of the sample. The normalized pore water pressure is equal to the pore pressure divided by the confining pressure, abbreviated as pore pressure. Figures 6.2 and 6.3, respectively, show the evolution over time of the drainage volumetric strain of the sample and the pore water pressure at four confining pressures (50 kPa, 100 kPa, 150 kPa, 200 kPa) and a reference temperature of 25 °C during the initial consolidation process. It can be observed that during the initial consolidation, the greater the consolidation pressure, the more drainage occurs, resulting in larger volumetric strains. The final total volumetric strains are 2.2%, 2.6%, 3.5%, and 4.8%, respectively. At the four confining pressures, the dissipation pattern of pore pressure is similar. Initially, the pore pressure dissipates rapidly as drainage begins, gradually decreasing until it reaches zero. For samples with higher confining pressures, the time taken for the pore pressure to stabilize is relatively longer.

After initial consolidation, samples with consolidation pressures of 200 kPa and 400 kPa were unloaded under undrained conditions to reduce the pressure to 50 kPa, followed by drainage consolidation. At this point, the overconsolidation ratios of the soil samples were 4 and 8, respectively. Figures 6.4 and 6.5 illustrate the changes in pore water pressure and volumetric strain during the unloading process. Unloading under undrained conditions causes the sample volume to expand, resulting in negative pore water pressure. The greater the consolidation pressure of the sample, the larger the amount of expansion and the larger the negative pore pressure (absolute value) generated, corresponding to the sample's expansion amount. After stabilizing the pore pressure and initiating drainage consolidation, the presence of negative pore pressure can cause a phenomenon known as "suction" in the sample [1]. The greater the negative pore pressure, the more significant the suction effect. However, relative

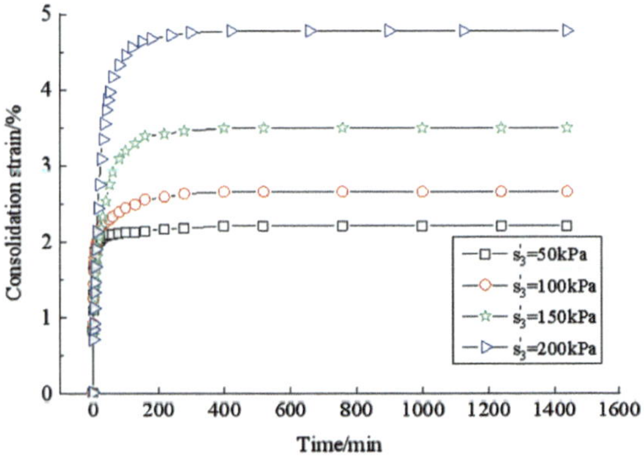

Fig. 6.2 Volumetric strain change during initial consolidation

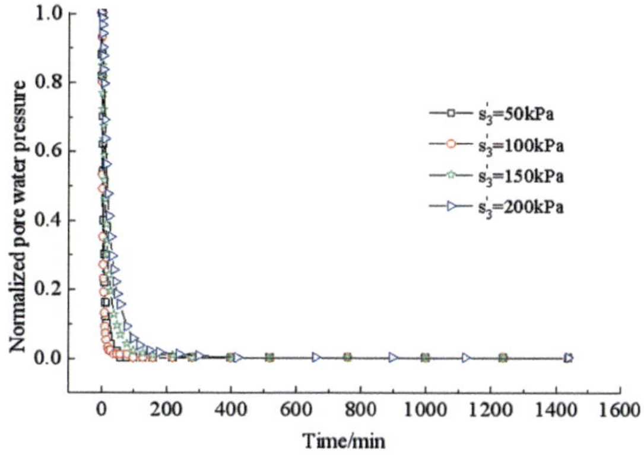

Fig. 6.3 Pore water pressure dissipation process during initial consolidation

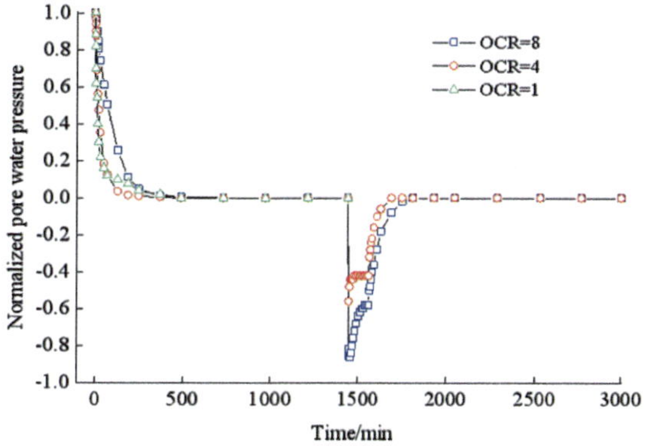

Fig. 6.4 Evolution of pore water pressure with time in primary consolidation process

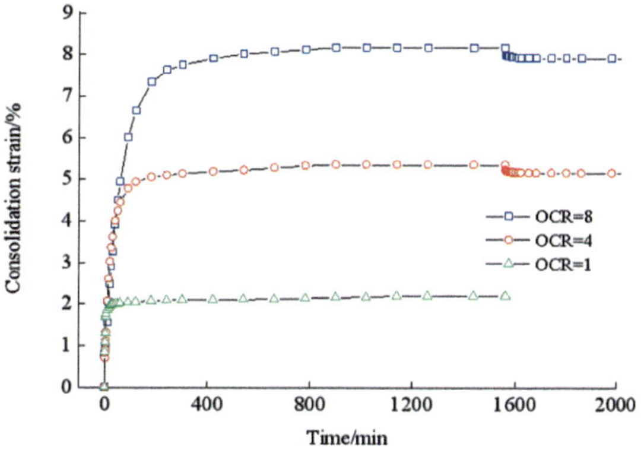

Fig. 6.5 Evolution of volumetric strains with tie in primary consolidation process

to the total drainage volume, the amount of water absorbed due to suction is relatively small. Therefore, after unloading, the overall strain in the sample only undergoes a slight reduction.

6.2.2 Thermal Consolidation Response Induced by Single-Step Heating–Cooling

Paths 1 and 2 in the experimental scheme represent temperature paths of a single-step heating–cooling process. Figure 6.6 illustrates the temperature variation process starting from room temperature of 25 °C, increasing to 50 and 75 °C, then decreasing back to room temperature of 25 °C. The heating power of the experimental setup is 500 W, with temperature measurement accuracy up to 0.1 °C, and minimal fluctuations when temperatures are stable. It can be observed that the heating/cooling rates are essentially consistent, with a temperature change rate of 0.5 °C/min during heating.

Figure 6.7 shows the variation of pore pressure over time during the process where different overconsolidation ratios of specimens are heated from 25 to 50 °C and 75 °C, and then cooled back to 25 °C. It can be observed that under undrained conditions, heating results in positive pore pressures, and the maximum pore pressure reached by specimens with different overconsolidation ratios is generally consistent. Cooling under undrained conditions generates negative pore water pressures, and there is an increasing trend in the maximum negative pore pressures generated by specimens with different overconsolidation ratios (more pronounced at 50 °C). Additionally, it is evident that larger temperature changes during undrained heating or cooling result in higher pore water pressures. The maximum normalized pore pressure reached during the 75 °C process (positive and negative) approaches 1, which is greater than the maximum pore pressure reached during the 50 °C process (0.6).

Burghignoli et al. [1] analysis indicates that the difference in thermal expansion coefficients between soil particles and pore water (where the thermal expansion coefficient of pore water is approximately 17 times that of soil particles) is the cause of positive or negative pore pressure generated during undrained heating or cooling processes. Due to the thermal expansion and contraction effects, during undrained heating, specimens tend to expand, with the expansion of pore water greater than that of soil particles. Under undrained conditions, where volume remains constant,

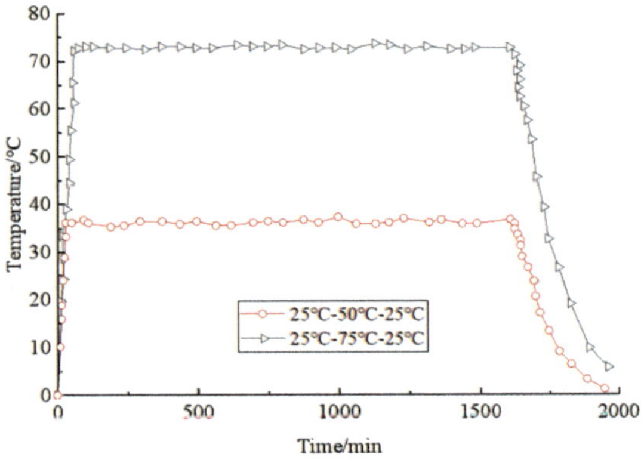

Fig. 6.6 The change process of test temperature

Fig. 6.7 Evolution of pore water pressure with time (single-stage heating–cooling process)

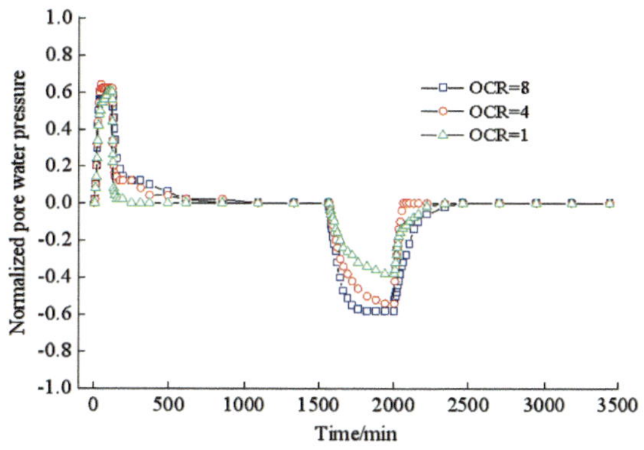

(a) 50°C

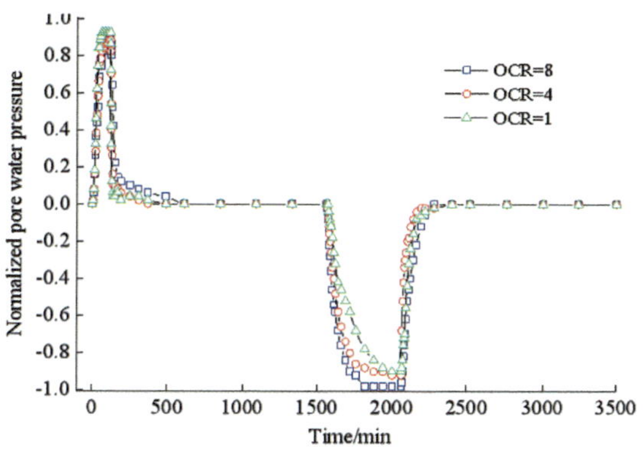

(b) 70°C

positive pore water pressure is required to "counteract" this expansion difference. Similarly, during undrained cooling, specimens tend to contract, with the contraction of pore water greater than that of soil particles. Under undrained conditions, negative pore water pressure is required to "counteract" this contraction difference. Specific pore pressure values generated by specimens with different overconsolidation ratios during undrained heating and cooling processes can be found in Table 6.3.

In Fig. 6.8, the evolution of volumetric strain during a single-stage heating–cooling process is shown for specimens with different overconsolidation ratios. The volumetric strain referred to in this experiment is the drained volumetric strain $\varepsilon_v = \Delta V/V_0$, which can be defined as the volumetric strain of the specimen after initial consolidation due to temperature changes (expressed as a percentage of the initial consolidation volume, V_0 represents the initial volume of the specimen after consolidation, and ΔV represents the consolidation displacement after heating/cooling). Upon heating and subsequent drainage consolidation, initially, a large amount of pore water is rapidly expelled, causing a rapid dissipation of

Table 6.3 Values of pore water pressure and volumetric strain caused by heating–cooling process

OCR	Heating–cooling path	u_{peak}/σ_3		$\varepsilon_v/\%$		General variation $\varepsilon_{vtot}/\%$
		Heating	Cooling	Heating consolidation	Cooling consolidation	
1	25 °C → 50 °C → 25 °C	0.60	− 0.38	1.03	0.54	0.49
1	25 °C → 75 °C → 25 °C	0.92	− 0.9	2.17	1.47	0.70
1	25 °C → 40 °C → 50 °C → 60 °C → 75 °C → 60 °C → 50 °C → 40 °C → 25 °C	0.46, 0.32, 0.34, 0.60	− 0.40, − 0.24, − 0.22, − 0.30	0.61,0.55, 0.71, 0.98	0.51, 0.47, 0.49, 0.49	0.90
4	25 °C → 50 °C → 25 °C	0.62	− 0.54	0.75	0.41	0.34
4	25 °C → 75 °C → 25 °C	0.88	− 0.92	1.55	0.91	0.64
4	25 °C → 40 °C → 50 °C → 60 °C → 75 °C → 60 °C → 50 °C → 40 °C → 25 °C	0.40, 0.36, 0.40, 0.62	− 0.44, − 0.30, − 0.30, − 0.40	0.55, 0.39, 0.43, 0.59	0.33, 0.16, 0.25, 0.36	0.86
8	25 °C → 50 °C → 25 °C	0.60	− 0.58	0.59	0.37	0.22
8	25 °C → 75 °C → 25 °C	0.87	− 0.98	1.29	0.76	0.53
8	25 °C → 40 °C → 50 °C → 60 °C → 75 °C → 60 °C → 50 °C → 40 °C → 25 °C	0.38, 0.34, 0.38, 0.60	− 0.46, − 0.34, − 0.34, − 0.42	0.,44, 0.28, 0.35, 0.56	0.28, 0.18, 0.22, 0.26	0.69

Fig. 6.8 Evolution of volumetric strains with time (single-stage heating–cooling process)

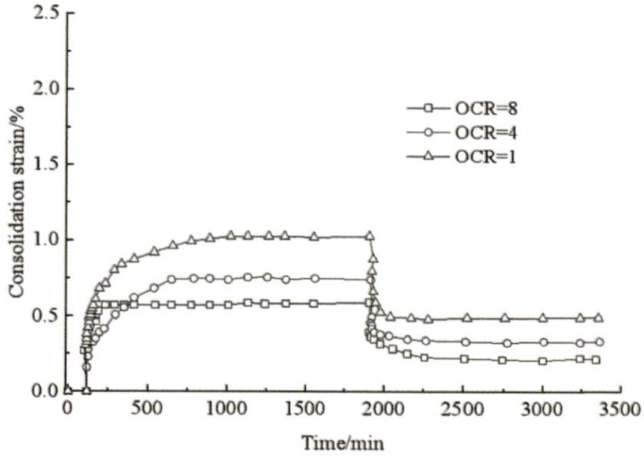

(a) 50°C

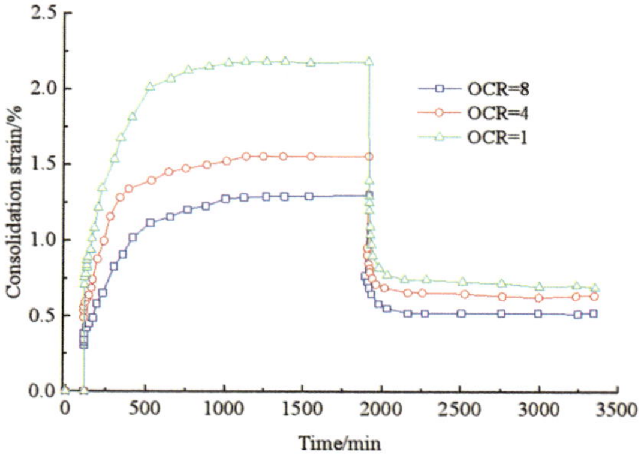

(b) 75°C

pore water pressure and an increase in volumetric strain. In the subsequent process, as pore pressure slowly dissipates, drainage volumetric strain increases gradually until pore pressure reaches zero, after which drainage volumetric strain stabilizes. It is evident that specimens with higher overconsolidation ratios exhibit smaller drainage volumetric strains after heating, corresponding to the thermal expansion observed in the introduction. Similar patterns are observed in drainage consolidation after cooling; negative pore pressures generated during cooling result in "suction" phenomena, causing specimen expansion. However, specimens with higher overconsolidation ratios experience slightly reduced water absorption. After a heating–cooling cycle, the total drainage volumetric strain of the specimen increases, indicating irreversible volumetric deformation, with specimens having higher overconsolidation ratios showing a trend of reduced total drainage volumetric strain. Additionally, larger heating–cooling amplitudes result in greater drainage (absorption) volumetric strains [2], with the drainage volumetric strain produced during 75 °C heating and cooling processes being greater than that during 50 °C processes, as detailed in Table 6.3.

As previously mentioned, the greater the temperature increase or decrease, the greater the induced pore water pressure, leading to larger strain in the drainage body caused by dissipation of pore pressure [1]. Furthermore, from a microscopic perspective, analysis shows that warming causes bound water around soil particles to transition to free water [3]. The greater the temperature change, the larger this transition, resulting in increased drainage volume. Additionally, warming reduces the viscosity of pore water, thereby increasing the permeability coefficient [3], which also contributes to increased drainage volume. Specimens with higher overconsolidation ratios have greater density and relatively less internal pore water content. Consequently, the strain in the drainage body after warming or cooling is relatively smaller, leading to a smaller overall strain.

6.2.3 Thermal Consolidation Response Due to Multiple Stage Heating–Cooling

In experimental pathway 3 of the test plan, a multi-stage heating–cooling process is employed, with drainage consolidation conducted after each stage of non-drained heating or cooling. Figure 6.9 illustrates the evolution over time of pore water pressure for different OCR (Over-Consolidation Ratio) samples during this multi-stage heating–cooling process. Similar to the single-stage heating/cooling patterns described earlier, larger temperature changes result in higher pore water pressure values. Samples with different OCRs achieve similar maximum normalized pore water pressures during each stage of heating or cooling, and larger OCR samples tend to exhibit increased negative pore pressures during the cooling process. For example, for an OCR of 4, samples heated stepwise from 25 °C to 40 °C, 50 °C, 60 °C, and 75 °C exhibit maximum normalized positive pore water pressures of 0.40, 0.36, 0.40, and 0.62, respectively. During stepwise cooling from 75 °C to 60 °C, 50 °C, 40 °C, and 25 °C, they reach maximum normalized negative pore water pressures of − 0.44, − 0.30, − 0.30, and − 0.40, respectively. For an OCR of 8, samples heated stepwise from 25 °C to 40 °C, 50 °C, 60 °C, and 75 °C exhibit maximum normalized positive pore water pressures of 0.38, 0.34, 0.38, and 0.60, respectively. During stepwise cooling from 75 °C to 60 °C, 50 °C, 40 °C, and 25 °C, they reach maximum normalized negative pore water pressures of − 0.46, − 0.34, − 0.34, and − 0.42, respectively. Table 6.3 provides specific values of the maximum normalized pore water pressures achieved during each stage of the multi-stage heating–cooling process for different samples with varying over-consolidation ratios.

Figure 6.10 depicts the variation in cumulative volumetric strain during a multi-stage heating–cooling process for samples with over-consolidation ratios (OCR) of 1, 4, and 8. The general trend is similar to that shown in Fig. 6.8 for single-stage heating–cooling processes: samples with higher over-consolidation ratios exhibit relatively smaller drained volumetric strains after heating or cooling, resulting in correspondingly smaller cumulative total volumetric strains or example, for an OCR

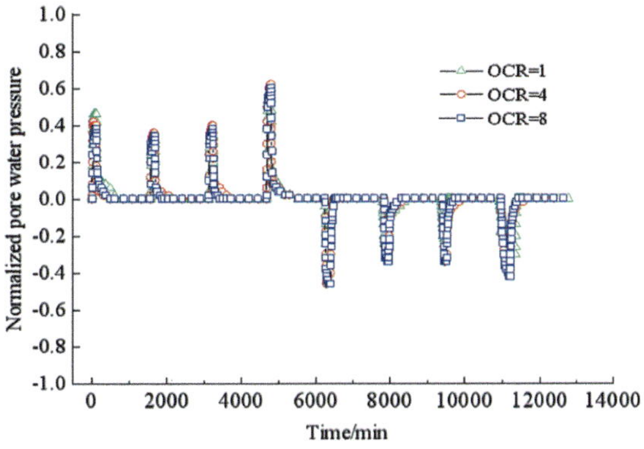

Fig. 6.9 Evolution of pore water with time (single-stage heating–cooling process)

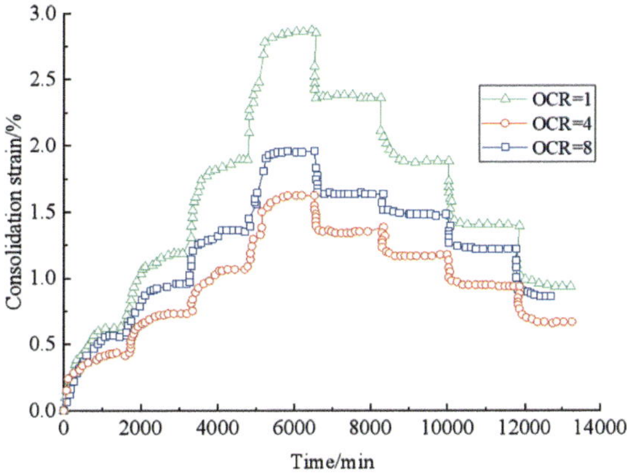

Fig. 6.10 Evolution of volumetric strains with time (isothermal multistage heating–cooling process)

of 4, samples heated stepwise from 25 to 75 °C show drained volumetric strains of 0.55, 0.39, 0.43, and 0.60% per stage, accumulating to a total heating volumetric strain of 1.96%. Subsequently, during stepwise cooling from 75 to 25 °C, the drained volumetric strains are 0.33, 0.16, 0.25, and 0.36% per stage, accumulating to a total cooling volumetric strain of 1.10%. Therefore, after one heating–cooling cycle, the total volumetric strain generated is 0.86%, indicating irreversible volume changes. Similar evolutionary patterns are observed for OCR = 1 and OCR = 8 samples, with specific values detailed in Table 6.3.

6.2.4 Comparative Analysis Under Different Temperature Paths

Figure 6.11 compares the evolution of total volumetric strain for samples with three different over-consolidation ratios (OCR) under single-stage and multi-stage

Fig. 6.11 Comparison of volumetric strains evolution between isothermal single-stage and multistage heating–cooling process for samples with different OCRs

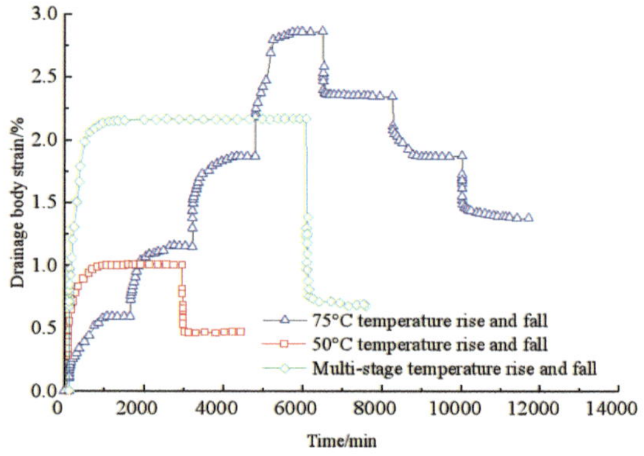

(a) OCR=1

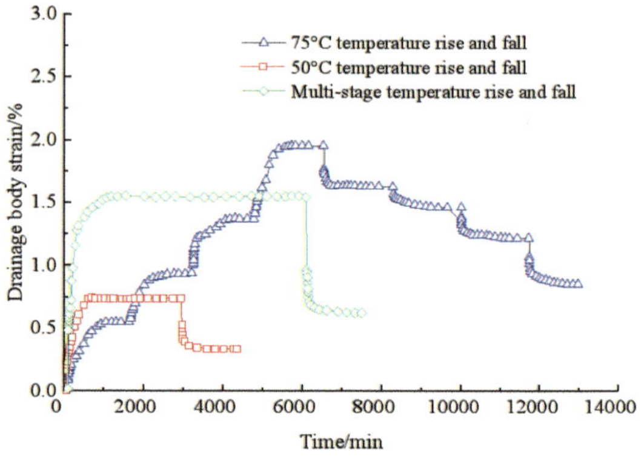

(b) OCR=4

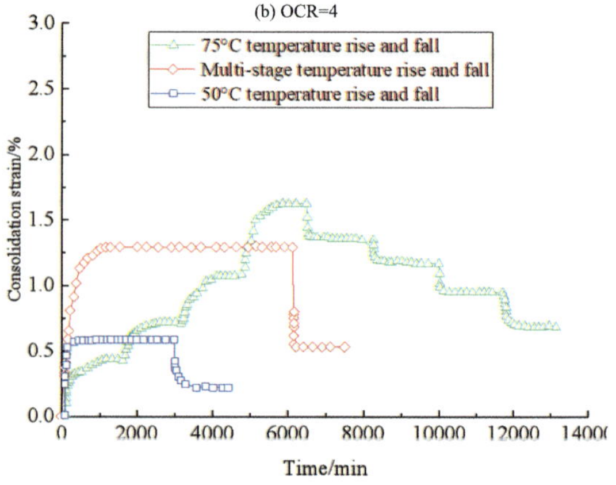

(c) OCR=8

heating–cooling processes. It can be observed that multi-stage processes result in larger total volumetric strains compared to single-stage processes. Additionally, the difference in strain between the two methods decreases with increasing over-consolidation ratio. For instance, during heating to 75 °C, the difference in total volumetric strain between multi-stage and single-stage processes for OCR = 1, 4, 8 samples is 0.68%, 0.41%, and 0.34%, respectively. Similarly, a similar pattern is observed during the cooling process. For example, during cooling from 75 °C to 25 °C, the difference in total volumetric strain between multi-stage and single-stage processes for OCR = 1, 4, 8 samples is 0.49%, 0.19%, and 0.16%, respectively. After one heating–cooling cycle, the total volumetric strain tends to be larger in multi-stage processes compared to single-stage processes, but the difference in total strain between multi-stage and single-stage processes remains relatively consistent across different over-consolidation ratios. For instance, for the 75 °C temperature process, the difference in total volumetric strain between multi-stage and single-stage processes for OCR = 1, 4, 8 samples is 0.20%, 0.22%, and 0.18%, respectively. The difference in total volumetric strain between multi-stage and single-stage processes can be attributed to two main factors: (1) With the same temperature range, the greater number of intermediate temperature levels in multi-stage processes results in a higher cumulative increase in pore pressure during non-drained heating. This leads to greater dissipation of pore pressure during drainage, resulting in higher volumetric strains. This is the primary reason for the observed differences. (2) Each stage in multi-stage processes includes 24 h of drainage time, which extends the consolidation period compared to single-stage processes. Time effects (creep) also contribute to this difference, although their impact is relatively minor [1, 3].

Table 6.3 lists the specific values of maximum pore water pressure and volumetric strain reached by samples with different over-consolidation ratios under various heating–cooling temperature paths. Based on this data, the analysis examines how volumetric strain varies with over-consolidation ratio across different temperature paths. Figure 6.12 illustrates the evolution of volumetric strain with over-consolidation ratio during heating and cooling processes at 50 and 75 °C, with the temperature paths detailed in Table 6.4. Figure 6.12 summarizes all the observed patterns in volumetric strain variations under different temperature paths: (1) Greater temperature increments (heating or cooling) result in larger volumetric strains. (2) Volumetric strains produced during heating or cooling processes, as well as the total volumetric strain after one heating–cooling cycle, decrease with increasing over-consolidation ratio, as previously discussed (see Fig. 6.8). (3) For heating or cooling processes to a specified temperature, the more intermediate temperature levels involved, the greater the resulting volumetric strain. For instance, multi-stage heating–cooling to 50 °C > multi-stage heating to 50 °C > single-stage heating to 50 °C.

Furthermore, it can be observed that the volumetric strain generated during the heating or cooling processes along different temperature paths in Fig. 6.12 exhibits an exponential relationship with the overconsolidation ratio (OCR), as indicated by the solid line in the figure. The relationship between volumetric strain and OCR can be expressed as:

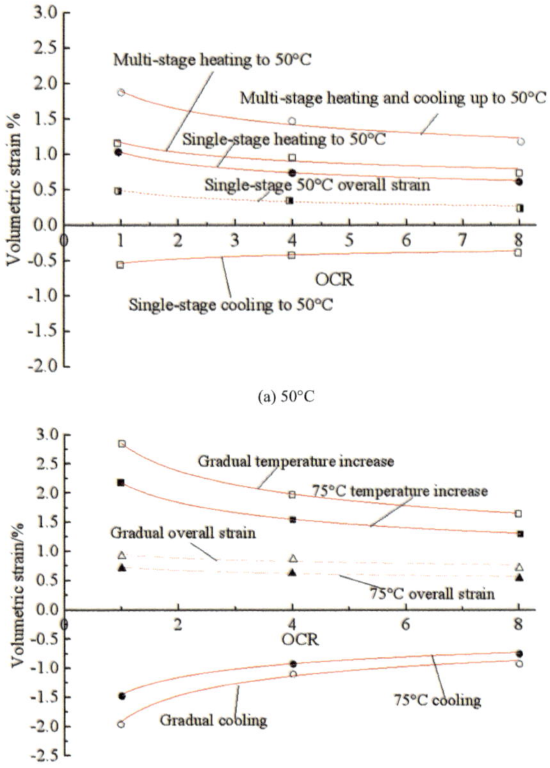

(a) 50°C

(b) 75°C

Fig. 6.12 Volumetric strain evolution with OCR under different heating–cooling paths and the fitting curves

$$\varepsilon_v = a(\text{OCR})^b \tag{6.7.1}$$

ε_v represents the volumetric strain generated during the heating or cooling process. Where a and b are parameters related to the temperature path. The dashed line in Fig. 6.12 represents the evolution of total volumetric strain with respect to the overconsolidation ratio (OCR) following a heating–cooling cycle, showing an approximate linear decreasing trend. Table 6.5 provides the exponential function fitting parameters for various heating/cooling cycles in relation to OCR. The R^2 value indicates the goodness of fit, ranging from 0 to 1, where values closer to 1 indicate a better fit. From the R^2 values, it can be observed that the relationship between volumetric strain generated under different temperature paths and OCR can be well approximated using exponential functions.

Table 6.4 Temperature paths

50 °C	Temperature path	75 °C	Temperature path
Single-stage heating	25 °C → 50 °C	Single-stage heating	25 °C → 75 °C
Single-stage heating	25 °C → 50 °C → 25 °C	Single-stage overall transformation	25 °C → 75 °C → 25 °C
Multi-stage heating	25 °C → 40 °C → 50 °C	Multi-stage heating	25 °C → 40 °C → 50 °C → 60 °C → 75 °C
Single-stage cooling	50 °C → 25 °C	Single-stage cooling	75 °C → 25 °C
Multi-stage heating and cooling	25 °C → 40 °C → 50 °C → 60 °C → 75 °C → 60 °C → 50 °C	Multi-stage cooling	75 °C → 60 °C → 50 °C → 40 °C → 25 °C
		Multi-stage overall transformation	25 °C → 40 °C → 50 °C → 60 °C → 75 °C → 60 °C → 50 °C → 40 °C → 25 °C

Table 6.5 Fitting parameters

Temperature path	*a*	*b*	R^2	Temperature path	*a*	*b*	R^2
Single-stage temperature increase of 50 °C	1.036	–0.256	0.981	Single-stage temperature increase of 75 °C	2.171	–0.248	0.999
Multi-stage temperature increase of 50 °C	1.174	–0.206	0.886	Multi-stage temperature increase of 75 °C	2.849	–0.269	0.999
Multi-stage temperature cycling of 50 °C	1.885	–0.210	0.949	Single-stage cooling of 75 °C	–1.469	–0.329	0.994
Single-stage cooling of 50 °C	–0.538	–0.186	0.992	Multi-stage cooling	–1.943	–0.374	0.982

6.3 Experimental Results Analysis—Method 2

6.3.1 Initial Consolidation Process

(1) Analysis of pore water pressure changes

1) OCR = 2

The experiments in this study used an overconsolidation ratio (OCR) of 2, with confining pressures set at four levels: 100–50 kPa, 200–100 kPa, 300–150 kPa, and 400–200 kPa. The consolidation duration for samples under each confining pressure was 1320 min. Taking the example of 100–50 kPa, after consolidating under a superconsolidated load of 100 kPa for 1320 min, the confining pressure was then reduced to 50 kPa under undrained conditions. The sample was then subjected to an additional 1320 min of consolidation under this 50 kPa confining pressure. After two consolidation cycles, soil samples were formed with an overconsolidation ratio (OCR) of 2. The consolidation processes under other confining pressure schemes were identical. The dissipation curves of pore water pressure during each sample's consolidation process are shown in Fig. 6.13.

Four sets of different confining pressure conditions were established for samples with an overconsolidation ratio (OCR) of 2. From the graph, it can be observed that the pore water pressure dissipation patterns are generally consistent across all four sets of samples. During the consolidation stage under the effect of superconsolidation load, at the beginning of consolidation, the ratio of pore water pressure to confining pressure (u/σ_3) for all samples was above 0.9. The u/σ_3 curve decreased rapidly initially, showing nearly linear behavior. Subsequently, the rate of decrease in u/σ_3 gradually slowed down, and eventually the curve tended to flatten. Under undrained conditions during pressure relief, the pore water pressure of the sample gradually decreases. The pore water pressures after relief to 50 kPa, 100 kPa, 150 kPa, and 200 kPa are – 12, – 21, – 25, and – 26 respectively, with corresponding u/σ_3 values

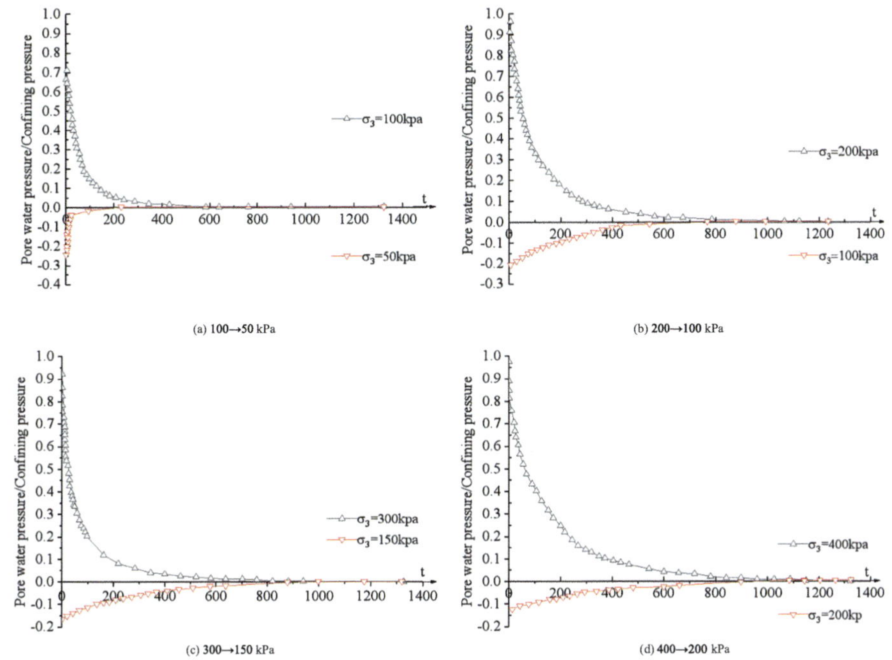

Fig. 6.13 Dissipation of pore pressure with time in consolidation processes under normal temperature (OCR = 2)

of – 0.24, – 0.21, – 0.17, and – 0.13. From the above data, it can be observed that as the confining pressure increases, the absolute value of negative pore water pressure induced by pressure relief increases. The absolute value of u/σ_3 decreases gradually. With time, the pore water pressure slowly rebounds and eventually returns to zero.

2) OCR = 3

The selected confining pressure values for this group of experiments are 150 → 50 kPa, 300 → 100 kPa, and 450 → 150 kPa. The consolidation process of the samples is the same as described earlier for OCR = 2. After two consolidation cycles of the samples, soil samples with an overconsolidation ratio (OCR) of 3 were formed. The pore water pressure dissipation curves during the consolidation process of each sample are shown in Fig. 6.14.

From the graph, it can be seen that the pore water pressure dissipation pattern of the samples is consistent with what was described earlier. As time progresses, the rate of decrease in u/σ_3 gradually diminishes, and the curve tends to flatten, indicating complete dissipation of pore water pressure.

Comparing the u/σ_3 variation curves under different confining pressure conditions, it is observed that with lower confining pressures, the u/σ_3 curve decreases more rapidly, and the time for the curve to flatten out is earlier. When the confining pressure

Fig. 6.14 Dissipation of pore pressure with time in consolidation processes under normal temperature (OCR = 3)

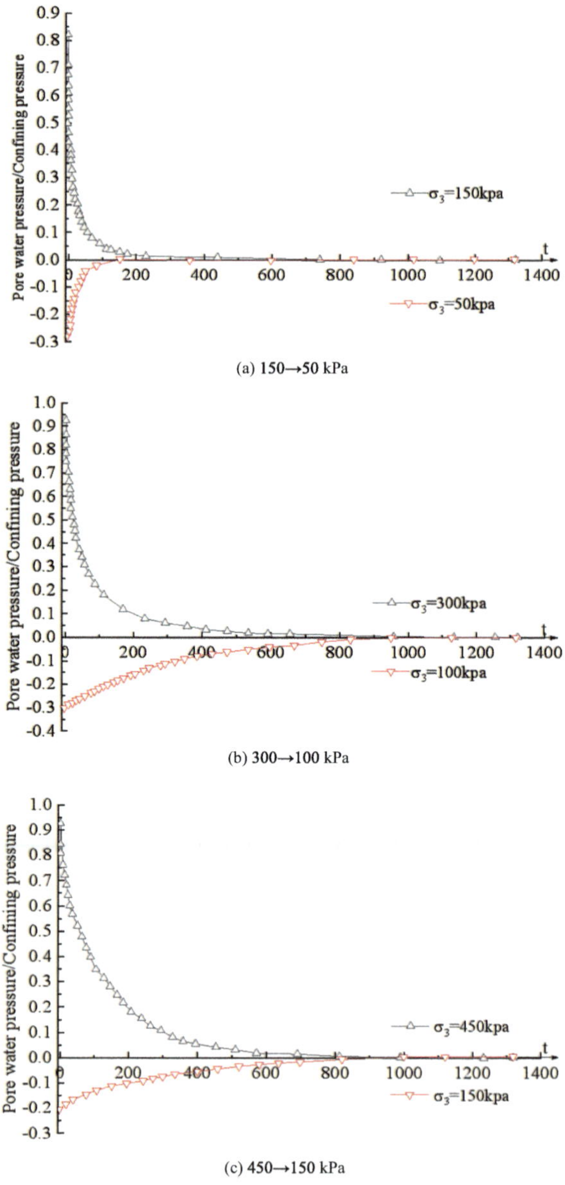

conditions were set at 150 → 50 kPa, the pore water pressure dissipates quickly, stabilizing after about 400 min. In contrast, with confining pressures of 450 → 150 kPa, the curve of u/σ_3 variation over time is noticeably smoother compared to other confining pressure conditions, requiring a longer time to stabilize, approximately around 1000 min.

3) OCR = 4

According to the experimental plan requirements, two sets of different confining pressure values were applied to the samples: 200 → 50 kPa and 400 → 100 kPa. Following the previously described steps for testing overconsolidated samples, the soil samples were consolidated accordingly. After stabilization, soil samples with an overconsolidation ratio (OCR) of 4 were formed. The pore water pressure dissipation curves during the consolidation process of each sample are shown in Fig. 6.15.

From the graph, it can be seen that the variation curve of pore water pressure to confining pressure ratio (u/σ_3) of the samples follows the same pattern as the previous two groups of samples. Due to the higher overconsolidation ratio in this group of samples, compared to the previous two groups, the absolute values of negative pore water pressure induced after unloading are larger. At confining pressures of 50 kPa and 100 kPa, they are − 21 and − 26, respectively. The ratio of pore water pressure to confining pressure (u/σ_3) is − 0.42 and − 0.26, respectively. Compared to samples with an overconsolidation ratio (OCR) of 2, the absolute values of u/σ_3 ratio after unloading are significantly larger in this group of samples, especially when the

Fig. 6.15 Dissipation of pore pressure with time in consolidation processes under normal temperature (OCR = 4)

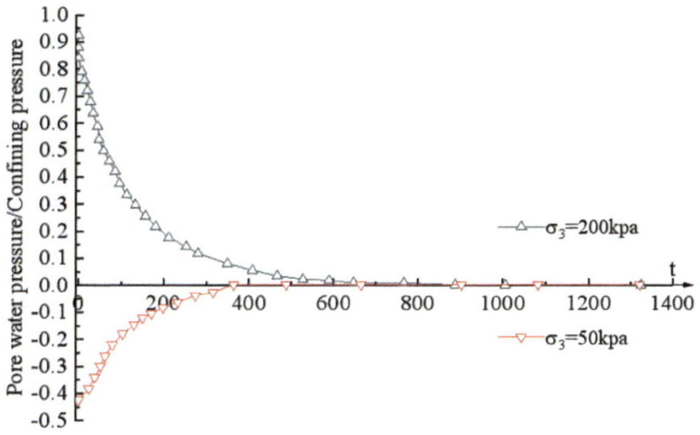

(a) 200→50 kPa

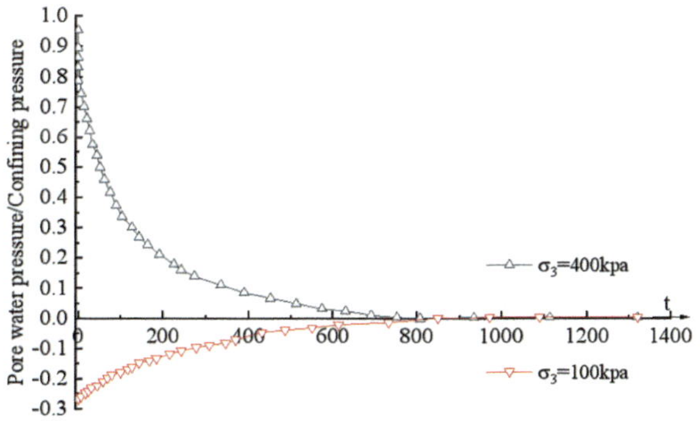

(b) 400→100 kPa

confining pressure is 50 kPa, where the u/σ_3 ratio is nearly twice that of the OCR = 2 samples.

From the pore water pressure dissipation curves during the consolidation processes of samples in Figs. 6.13, 6.14 and 6.15, it can be observed that with increasing consolidation time, the rate of evolution of the u/σ_3 curve gradually decreases. The time required for pore water pressure to decrease by the same amount also increases. When u/σ_3 drops below 0.05, the curve becomes flat, indicating very slow pore water pressure dissipation. Comparing the pore water pressure dissipation curves under different confining pressure conditions during sample consolidation, it is found that the evolution rate of the u/σ_3 curve slows down with increasing confining pressure. At the same time, as the applied confining pressure increases, the required consolidation time for the samples also increases. With an increase in overconsolidation ratio, the absolute value of negative pore water pressure induced after unloading stabilizes also increases. Correspondingly, the absolute value of the ratio of pore water pressure to confining pressure (u/σ_3) also increases.

(2) Changes in volumetric strain

The volumetric strain εv variation curve of the samples is shown in Fig. 6.16. The curve depicts the evolution of volumetric strain during the consolidation stage under overconsolidated loading for the initial 1320 min. From the graph, it can be observed that under different confining pressure conditions (100, 200, 300, 400 kPa), the evolution curves of volumetric strain ε_v for the samples are similar, showing a consistent trend in volumetric strain variation. At the beginning of consolidation, there is a significant instantaneous increment in volumetric strain. During the early stages of consolidation, the rate of volumetric strain growth is relatively fast, with the change in volumetric strain exceeding 60% of the total volumetric strain at completion within the first 20 min. Subsequently, the rate of volumetric strain increase gradually slows down, and the curve becomes increasingly flat. Eventually, the volumetric strain value remains relatively unchanged for a long period, indicating completion of consolidation under the influence of overconsolidated loading.

After the aforementioned consolidation stage concluded, according to experimental requirements, partial confining pressure was released from the sample under undrained conditions to a stable state, marking the beginning of the subsequent consolidation stage. The evolution of volumetric strain during this consolidation process is shown in the following figure. From the graph, it can be seen that after unloading, the volumetric strain during this consolidation stage gradually decreases. Due to the reduction in confining pressure, the sample exhibits swelling consolidation. During this stage, the evolution trend of consolidation volumetric strain is similar to the previous stage, with the rate of change gradually decreasing. Eventually, the curve stabilizes, indicating that consolidation of the sample is essentially complete. Compared to the volumetric strain changes under overconsolidated loading, the decrease in volumetric strain during this stage is minimal. After completion of both stages of consolidation, the total volumetric strain of the sample becomes positive, indicating volume contraction of the sample.

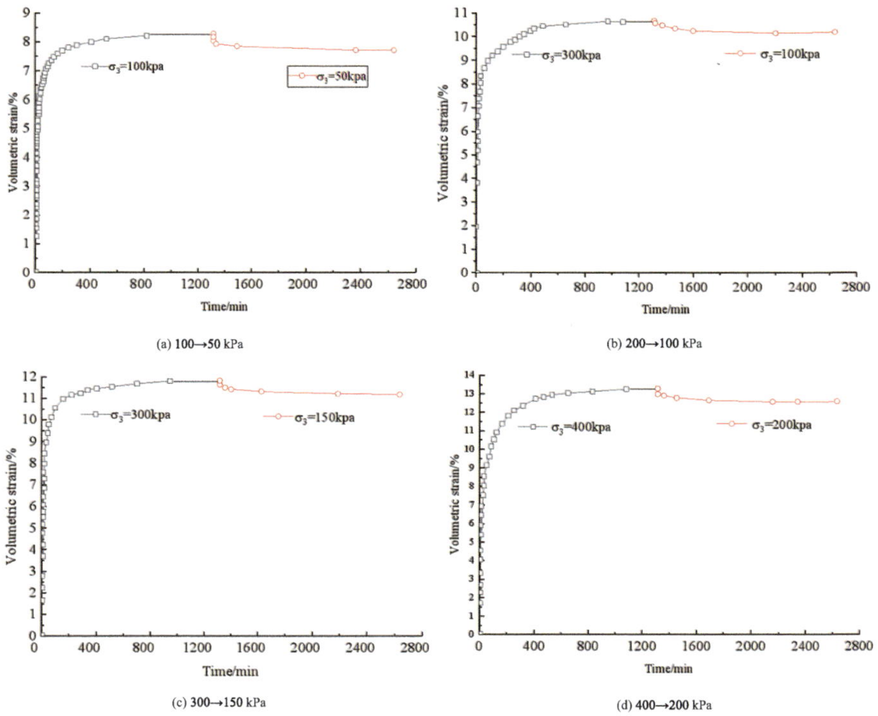

Fig. 6.16 Volumetric strain pressure with time consolidation under normal temperature (OCR = 2)

Comparing four sets of experimental data, it was found that as the confining pressure applied to the samples increased, there was a noticeable increase in the consolidation volumetric strain ε_{v1} under overconsolidated loading. After unloading, the absolute change in volumetric strain $|\varepsilon_{v2}|$ during the consolidation stage also showed a slight increase, but the magnitude of $|\varepsilon_{v2}|$ was smaller than the increase in ε_{v1}. The volumetric strain values of the four sets of samples increased with the rise in confining pressure levels.

2) OCR = 3

Figure 6.17 presents the evolution curves of volumetric strain ε_v during the consolidation process of samples under different confining pressures, with an overconsolidation ratio (OCR) of 3. The dissipation process of pore water pressure during the corresponding consolidation stages was discussed in Sect. 4.1.1, and the experimental methods and consolidation times are consistent with those described earlier.

From the curves in Fig. 6.17, it can be observed that under the three different confining pressure conditions chosen for the experiments, the evolution curves of volumetric strain ε_v for the samples exhibit similar patterns. Moreover, the change trends of volumetric strain curves during the consolidation process for each group

Fig. 6.17 Volumetric strain with time in consolidation processes under normal temperature (OCR = 3)

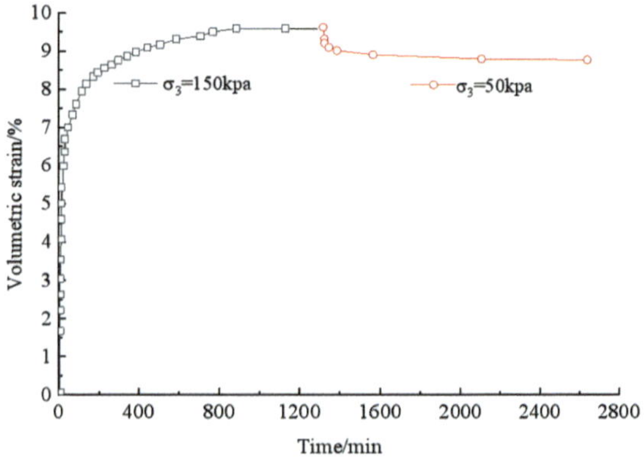

(a) 150→50 kPa

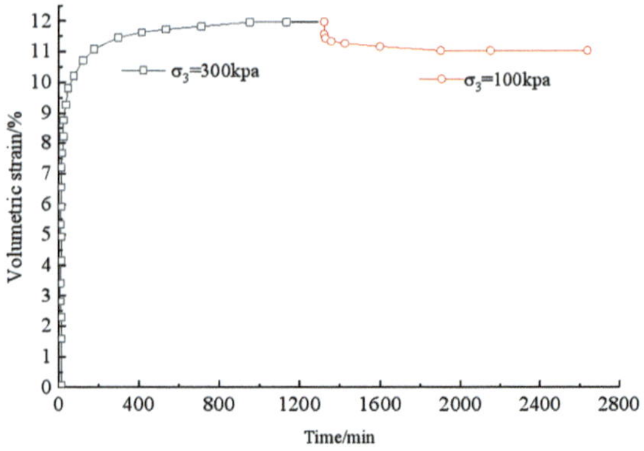

(b) 300→100 kPa

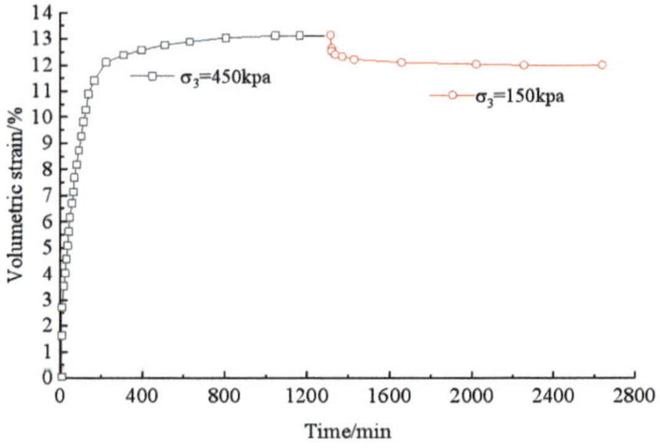

(c) 450→150 kPa

of samples are consistent with the earlier analysis of the sample with an overconsolidation ratio (OCR) of 2. The rates of volumetric strain change during the two consolidation stages gradually decrease with increasing time, and the curves tend to flatten, indicating that the samples undergo consolidation at specific confining pressure levels.

Analyze and compare the consolidation volumetric strains of the three samples under different confining pressure conditions upon completion of consolidation under overconsolidated loading. The volumetric strains ε_{v1} after consolidation for samples S9 to S11 are 9.58%, 11.97%, and 13.12%, respectively, showing an increase with higher confining pressures. After unloading consolidation, the volumetric strains ε_{v2} decrease, with values of − 0.83%, − 0.94%, and − 1.15% for these three samples, respectively. In the unloading consolidation phase under the same overconsolidation ratio conditions, the volumetric strains ε_{v2} increase with higher confining pressure levels. Based on the data above, the difference $|\Delta\varepsilon_{v2}|$ between consolidation strains among different samples is less than $|\Delta\varepsilon_{v1}|$, indicating that the total volumetric strains after two consolidations also increase with higher confining pressures. The initial volumetric strains after completion of consolidation for these three samples are 8.74%, 11.04%, and 11.97%, respectively.

3) OCR = 4

Figure 6.18 shows the evolution curves of consolidation volumetric strain ε_v over time for two samples under different confining pressure conditions with an overconsolidation ratio (OCR) of 4. The previous sections have analyzed and discussed the evolution patterns of volumetric strain during consolidation for two groups of samples with OCRs of 2 and 3, respectively. From Figs. 6.16, 6.17 and 6.18, it can be observed that the trends in volumetric strain during consolidation for the samples in this group are similar to those in the previous two groups. Due to the higher overconsolidation ratio, the volumetric strains generated after unloading consolidation under the same confining pressure conditions differ from those of the previous two groups.

Comparing Fig. 6.18 a and b, it can be observed that: during the initial 1320 min of consolidation (Stage C1 under overconsolidated loading), samples under lower confining pressures exhibit faster changes in volumetric strain, and their curves stabilize earlier. This indicates that samples require less time to complete consolidation under lower confining pressures, consistent with conclusions drawn from the analysis of pore water pressure variations in the previous section. Similarly, Stage C2 during unloading consolidation shows a similar pattern.

The volumetric strains produced after completion of consolidation in stages C1 and C2 increase numerically with higher confining pressures. Given the larger overconsolidation ratio selected for this group, compared to the previous two groups of samples, under the same confining pressure conditions after unloading, the volumetric strains during the suction consolidation stage for this group show a slight increase.

The consolidation volumetric strains during various stages of initial consolidation and the total volumetric strains after consolidation completion for the three groups

Fig. 6.18 Volumetric strain pressure with time in consolidation processes under normal temperature (OCR = 4)

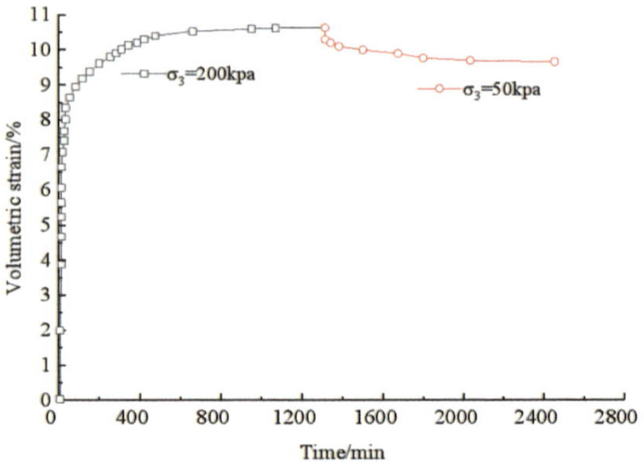

(a) 200→50 kPa

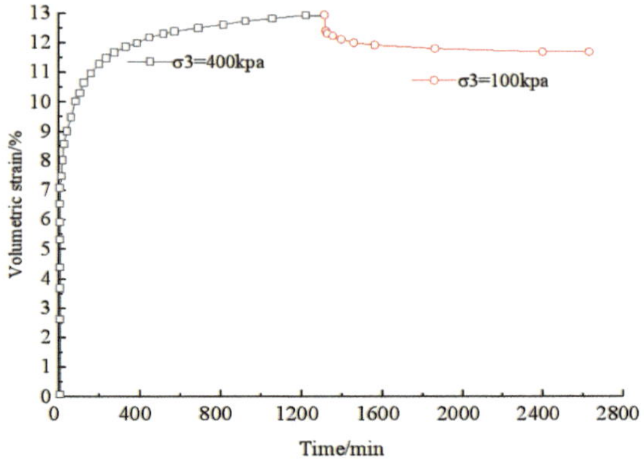

(b) 400→100 kPa

of samples with different overconsolidation ratios are shown in Table 6.6. From the data in the table, it can be observed that at a confining pressure of $\sigma_3 = 50$ kPa, the volumetric strains during stage C2 for samples with different overconsolidation ratios are – 0.52%, – 0.83%, and – 0.94%, respectively. At a confining pressure of $\sigma_3 = 100$ kPa, the volumetric strains during stage C2 for these samples are – 0.52%, – 0.94%, and – 1.25%, respectively. This indicates that under the same confining pressure, due to different overconsolidation loads, the volumetric strains after unloading consolidation vary, and their values increase with higher overconsolidation ratios. Regarding the change in final total volumetric strain after initial consolidation completion, the effect of different overconsolidation loads is also significant. At $\sigma_3 = 50$ kPa, the final total volumetric strains for samples with different overconsolidation ratios are 7.70%, 8.74%, and 9.68%, respectively. At $\sigma_3 = 100$ kPa, the final total volumetric strains are 10.10%, 11.04%, and 11.66%, respectively. This shows that under the same confining pressure conditions, the final total volumetric strains of samples also increase with higher overconsolidation ratios.

Table 6.6 Volumetric strain of overconsolidated clay samples

OCR	Confining pressure (kPa)	Overconsolidation load (kPa)	Volumetric strain during stage C1 (%)	Volumetric strain during stage C2 (%)	Final volumetric strain (%)
2	50	100	8.22	− 0.52	7.70
	100	200	10.62	− 0.52	10.10
	150	300	11.76	− 0.62	11.14
	200	400	13.22	− 0.73	12.49
3	50	150	9.58	− 0.83	8.74
	100	300	11.97	− 0.94	11.04
	150	450	13.12	− 1.15	11.97
4	50	200	10.62	− 0.94	9.68
	100	400	12.91	− 1.25	11.66

When comparing the data in Table 6.6, it is observed that the volumetric strain evolution curves and final volumetric strains during the C1 stage under the same overconsolidation load differ slightly among different overconsolidation ratio groups. These differences arise due to constraints in experimental conditions and inevitable minor errors during operations. However, the impact of these subtle differences on the consolidation law and the analysis of test results can be considered negligible.

6.3.2 *Thermal Consolidation Response of Overconsolidated Soil Under Different Confining Pressures*

(1) Analysis of pore water pressure variations

1) OCR = 2

The four curves in Fig. 6.19 represent the evolution of pore water pressure for samples S5–S8 (OCR = 2) after completion of initial consolidation under thermal loading. The undrained heating and cooling stages lasted for 300 min each, and the isothermal drained consolidation time was 1140 min. Each cycle for the samples took 2880 min.

From the figure, it can be observed that under the same overconsolidation ratio conditions, samples at different confining pressure levels exhibit consistent trends in pore water pressure evolution during the temperature cycling process. During the heating phase, pore water pressure in the samples increases continuously with temperature rise, reaching its peak as the temperature reaches 75 °C. After the heating phase, the samples undergo drained consolidation at 75 °C, during which pore water pressure gradually dissipates as water is expelled. Initially, the dissipation rate is rapid, but it decreases with prolonged consolidation time, as indicated by the trend in the u/σ_3 ratio curve in the figure, which becomes more gradual until it flattens out,

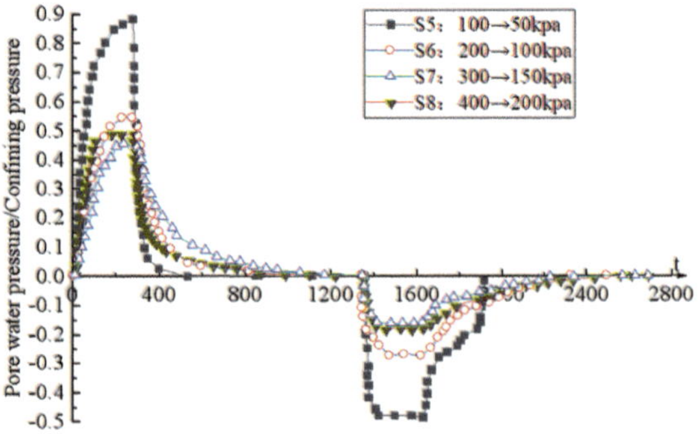

Fig. 6.19 Pore pressure evolutions under thermal load (OCR = 2)

indicating complete dissipation of pore water pressure. The pattern of the u/σ_3 ratio curve during the cooling and subsequent isothermal consolidation stages follows a similar pattern to the heating phase. Comparing the peak u/σ_3 ratio curves during the heating and cooling stages of the four curves, it is noted that sample S5 exhibits the highest peak u/σ_3 ratio. After stabilization during heating and cooling, the u/σ_3 ratios for sample S5 are 0.88 and − 0.48 respectively, significantly higher than those for the other three samples. Sample S6 shows a slightly higher peak u/σ_3 ratio compared to samples S7 and S8. When comparing samples S7 and S8, their peak u/σ_3 ratios are similar, although the evolution rate of pore water pressure during consolidation and the time required to achieve stabilization differ between the two samples.

During the heating process, the expansion of pore water due to temperature increase far exceeds the expansion of the soil particle skeleton. This differential expansion generates positive pore water pressures within the sample. After initial consolidation, the volumetric strains for samples S5 to S8 are 7.70%, 10.10%, 10.62%, and 12.49%, respectively. The greater the confining pressure applied to the sample during consolidation, the larger the amount of pore water expelled during initial consolidation, and consequently, the less residual pore water remains inside the sample upon completion of consolidation. Although the pore water pressure generated during heating increases with increasing confining pressure, the influence of confining pressure on pore water pressure is less significant compared to the expansion differential induced by temperature when there is a large difference in pore water volume inside the sample. Therefore, sample S5 exhibits a larger peak u/σ_3 ratio compared to the other three samples. For samples S6 to S8, after completion of initial consolidation, there is less residual pore water inside the samples. The influence of temperature may no longer have a clear advantage compared to the effect of confining pressure on increasing pore water pressure. Hence, the peak u/σ_3 ratios for samples S6 to S8 are relatively similar.

2) OCR = 3 and OCR = 4

The pore water pressure evolution curves during temperature loading for two sets of samples with overconsolidation ratios (OCR) of 3 and 4 are shown in Fig. 6.20. From the figure, it can be observed that the u/σ3 ratio curves for the OCR = 3 and OCR = 4 samples exhibit similar evolution patterns during temperature loading as previously analyzed for the OCR = 2 samples. The evolution process of pore water pressure and the reasons for differences in peak u/σ_3 ratios among samples with the same overconsolidation ratio have been detailed in the previous analysis and will not be reiterated here.

(2) Analysis of volumetric strain variations

1) OCR = 2

The volumetric strain (ε_v) variation curves under temperature loading for specimens S5–S8 (OCR = 2) are shown in Fig. 6.21. From the figure, it can be observed that the volumetric strain evolution curves during temperature cycling are consistent for specimens subjected to different confining pressures. During each cycle, in the heating and cooling phases, the specimens remain undrained, resulting in no change in volumetric strain during these stages. After heating, positive pore water pressure is generated due to pore water expansion. Upon the initiation of drainage consolidation, as pore water pressure dissipates and water is expelled, the specimen

Fig. 6.20 Pore pressure evolutions under thermal load

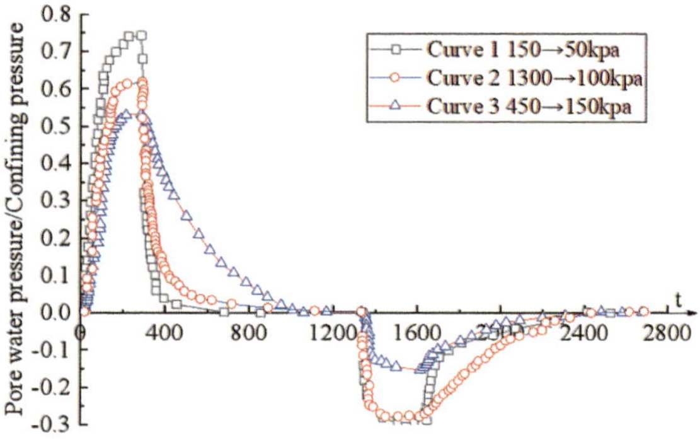

(a) OCR=3

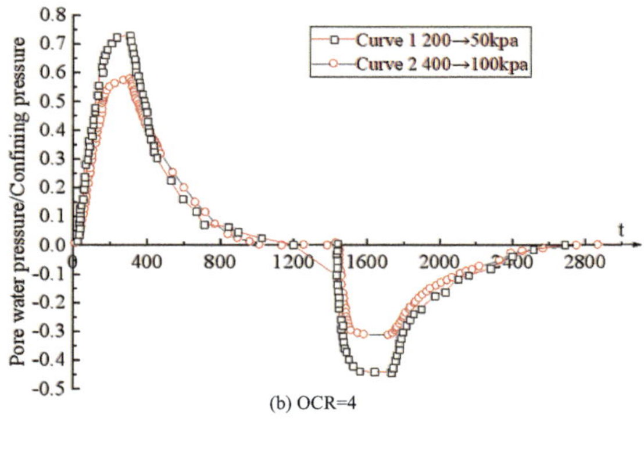

(b) OCR=4

(b) OCR=4

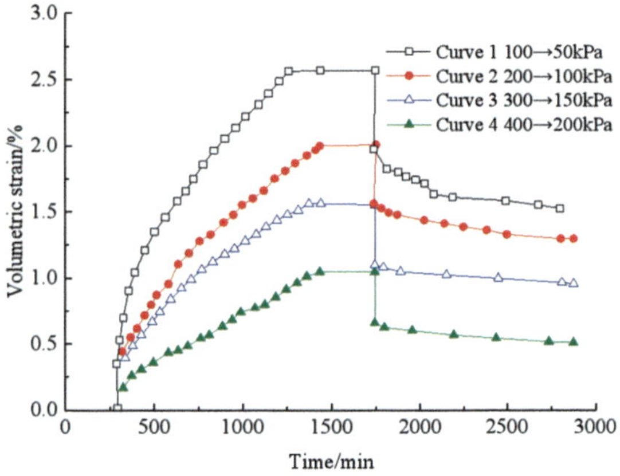

Fig. 6.21 Volumetric strain with time under thermal load (OCR = 2)

contracts, leading to a gradual increase in volumetric strain. During cooling, pore water contracts, generating negative pore water pressure inside the specimen. Subsequently, during the consolidation phase, the specimen exhibits water absorption, increasing pore water, causing expansion and resulting in negative volumetric strain during this phase. Overall, the volumetric strain gradually decreases from the value achieved upon completion of the consolidation phase of the previous cycle.

Comparing the two consolidation stages at different temperatures, it is evident that the volumetric contraction of the specimen during isothermal consolidation at 75 °C is greater than the volumetric expansion during consolidation at 25 °C. Therefore, after one cycle is completed, the total volumetric strain of the specimen is positive, indicating shrinkage in soil deformation. Comparing the volumetric strain evolution curves of four specimens under different confining pressures as shown in the figure, it is found that the higher the confining pressure applied to the specimen, the smaller the volumetric strain during isothermal consolidation at 75 °C. After completing one temperature cycle, the final volumetric strain value of the specimen is also smaller. Combining the previous section's analysis of pore water pressure evolution under temperature loading for specimens S5–S8, it can be inferred that the magnitude of volumetric strain during consolidation under thermal-hydro-mechanical coupling is related to the ratio of pore water pressure induced by temperature changes inside the specimen to the confining pressure (u/σ_3). A higher u/σ_3 ratio results in a larger consolidation volumetric strain value for the specimen.

2) OCR = 3 and OCR = 4

Figure 6.22 depicts the volumetric strain (ε_v) versus time curves during temperature loading for two groups of specimens with overconsolidation ratios (OCR) of 3 and 4, respectively. From the figure, it can be observed that the volumetric strain evolution during temperature loading is consistent with the description above. Additionally, when specimens have the same OCR, the peak values of the ε_v versus time curves decrease with increasing confining pressure, and the volumetric strain values at the end of each cycle exhibit a similar pattern.

Fig. 6.22 Volumetric strain with under thermal load

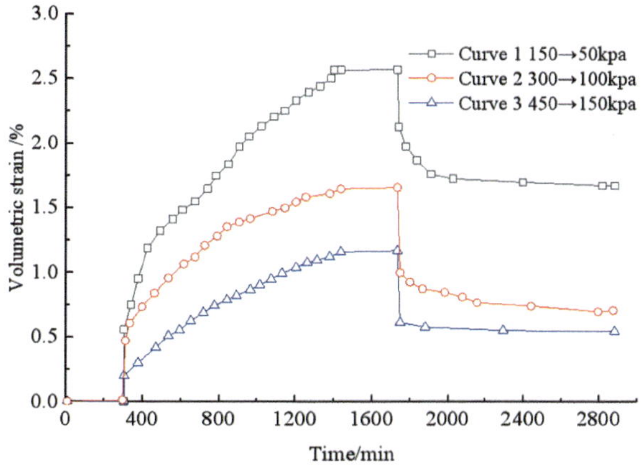

(a) OCR=3

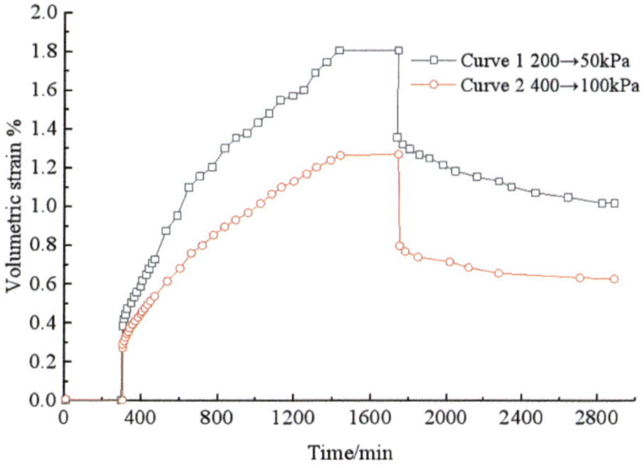

(b) OCR=4

6.4 Summary

(1) The greater the magnitude of undrained heating or cooling, the larger the induced pore water pressure. Samples with different overconsolidation ratios show similar maximum pore water pressures reached during heating, but during cooling, there is an increasing trend in the generation of negative pore pressures with higher overconsolidation ratios.
(2) The larger the magnitude of heating or cooling, the greater the consolidation volumetric strain produced. Additionally, consolidation volumetric strain decreases exponentially with increasing overconsolidation ratio. After one heating–cooling cycle, the total volumetric strain also decreases with increasing overconsolidation ratio.

(3) The more temperature levels set during the heating or cooling process, the greater the corresponding consolidation volumetric strain. Additionally, with increasing overconsolidation ratio, the difference in volumetric strain between different levels decreases.
(4) For multiple specimens subjected to different confining pressures, when the overconsolidation ratio (OCR) is the same, the dissipation pattern of pore water pressure during initial consolidation remains consistent. However, the consolidation rate and the time required to complete consolidation vary with different levels of confining pressure. When the confining pressure is lower, the u/σ_3 curve descends more rapidly, flattening earlier, and the consolidation process is shorter.
(5) For specimens with the same overconsolidation ratio (OCR), the evolution pattern of volumetric strain (ε_v) during initial consolidation is similar. As consolidation time increases, the rate of volumetric strain growth gradually decreases, and in the later stages of consolidation, the volumetric strain values stabilize without significant further changes, resulting in a flattened curve. Under higher overconsolidation loads, the final volumetric strain value of specimens increases with higher confining pressure levels.
(6) When the overconsolidation ratio (OCR) is the same, the peak value of the u/σ3 ratio curve during the heating and cooling stages decreases with increasing confining pressure. This decreasing trend is more pronounced at lower confining pressures, while under higher confining pressures, the difference in u/σ_3 peak values is not significant. After temperature increases, the volumetric strain variation during the isothermal consolidation stage of specimens is related to the magnitude of the u/σ_3 ratio; a larger u/σ_3 ratio corresponds to a greater change in consolidation volumetric strain. As the confining pressure increases, the peak value of the u/σ_3 ratio curve during the heating stage decreases. Therefore, the variation in volumetric strain during the isothermal consolidation process of specimens decreases with increasing confining pressure.

References

1. Burghignoli A, Desideri A, Miliziano S (2000) A laboratory study on the thermomechanical behaviour of clayey soils. Can Geotech J 37(4):764–780
2. Sultan N, Delage P, Cui YJ (2002) Temperature effects on the volume change behaviour of Boom clay. Eng Geol 64(2–3):135–145
3. Towhata I, Kuntiwattanaul P, Seko I, Ohishi K (1993) Volume change of clays induced by heating as observed in consolidation tests. Soils Found 33:170–183

Chapter 7
Thermal Consolidation Model of Saturated Clay Based on Thermoelastic Theory

7.1 State Equations

The material properties parameters of soil, such as density, porosity, dynamic viscosity, specific heat, compressibility, thermal conductivity, thermal expansion coefficient, and other state variables, all vary with changes in pressure and temperature, and they can all be expressed as functions of pressure and temperature. Let f be a certain state variable, then the general form of the state equation can be written as:

$$f = f(P, T) \tag{7.1}$$

Its total differential is:

$$\mathrm{d}f = \left(\frac{\partial f}{\partial P}\right)_{\mathrm{T}} \mathrm{d}P + \left(\frac{\partial f}{\partial T}\right)_{\mathrm{P}} \mathrm{d}T \tag{7.2}$$

$$\frac{\mathrm{d}f}{f} = \frac{1}{f}\left(\frac{\partial f}{\partial P}\right)_{\mathrm{T}} \mathrm{d}P + \frac{1}{f}\left(\frac{\partial f}{\partial T}\right)_{\mathrm{P}} \mathrm{d}T \tag{7.3}$$

When temperature is constant, the rate of change of f with respect to unit pressure increase is expressed as $\frac{1}{f}\left(\frac{\partial f}{\partial T}\right)_{\mathrm{P}}$. The rate of change of f with respect to unit temperature increase when pressure is constant. Both of these rates are typically not constants but functions of temperature and pressure.

In the three-field coupling study of this paper, the variables such as pore water density (p_{w}), soil skeleton density (p_{s}), viscosity of pore water (μ_{f}), porosity (ϕ), and permeability (k) are represented using the state variables P and T.

G. Yang and B. Bai, *Experimental and Theoretical Study on Thermal Consolidation Behavior of Saturated Clay*, https://doi.org/10.1007/978-981-96-6909-7_7

(1) Density and porosity

The equation of state for soil skeleton density can generally be written as:

$$\rho_s = \rho_{s0}\left[1 + \frac{P - P_0}{K_s} - 3\beta_s(T - T_0) + \frac{\overline{P'}}{(1-\phi)K_s}\right] \tag{7.4}$$

The equation of state for pore water density can generally be written as:

$$\rho_w = \rho_{w0}[1 + c_w(P - P_0) - \beta_w(T - T_0)] \tag{7.5}$$

The equation of state for porosity can be expressed as:

$$\phi = 1 - \frac{1-\phi_0}{1+\varepsilon_v}\left[1 - \frac{P - P_0}{K_s} + \beta_s(T - T_0)\right] \tag{7.6}$$

It can also generally be expressed as:

$$\phi = \phi_0\left[1 + c_\phi(P - P_0) + \beta_\phi(T - T_0)\right] \tag{7.7}$$

According to Eq. (7.4), taking the derivative with respect to (t) yields the following expression:

$$\frac{(1-\phi)}{\rho_s}\frac{\partial \rho_s}{\partial t} = -\frac{K_b}{K_s}\frac{\partial \varepsilon_v}{\partial t} + \frac{1}{K_s}\left[(1-\phi) - \frac{K_b}{K_s}\right]\frac{\partial P}{\partial t} - 3\left[(1-\phi)\beta_s - \frac{K_b}{K_s}\beta_b\right]\frac{\partial T}{\partial t} \tag{7.8}$$

The overall bulk modulus (K_b); The overall linear expansion coefficient (β_b), $3\beta_s = \frac{1}{\rho_s}\frac{\partial \rho_s}{\partial T}$, $3\beta_b = \frac{1}{V_b}\frac{\partial V_b}{\partial T}$, ρ_{s0}, ρ_{w0}, ϕ_0 is the pressure $P = P_0$, temperature $T = T_0$, the densities of the soil skeleton, pore water, and porosity; β_s, β_w, β_f the volumetric thermal expansion coefficients for the soil skeleton, pore water, and porosity, respectively; the bulk modulus of the soil skeleton (K_s); the average effective stress (P'); the volumetric compressibility coefficients for pore water and porosity, respectively, (c_w, c_ϕ) volumetric strain (ε_v).

(2) The viscosity of water (μ_f)

It changes very little with pressure and can be neglected, but it varies significantly with temperature, from 1.794 mPas at 0 °C to 0.284 mPas at 100 °C, a difference of about six times. This can be expressed as:

$$\mu_f = 1.984 \times 10^{-6} \exp\left(\frac{1825.85}{T}\right)(\mathrm{N\,s/m^2}), \tag{7.9}$$

where T is the temperature in Kelvin, $T = {}^\circ\mathrm{C} + 273.15\ {}^\circ\mathrm{C}$.

(3) Permeability k

There are many expressions for permeability k. During the coupled seepage process, it is constrained by various factors and changes continuously. According to the seepage mechanics Kozeny equation and relevant literature, it can be derived that the equation for permeability k is:

$$k = \frac{k_0}{1+\varepsilon_v}\left[1 + \frac{\varepsilon_v}{\phi_0} - \frac{(\beta_s \Delta T + (P - P_0)/K_s)(1-\phi_0)}{\phi_0}\right]^3 \tag{7.10}$$

7.2 Constitutive Equations

The constitutive equation describes the inherent response that depends on the internal structure of a specific material. It reflects the thermal motion characteristics of the material's molecules or the intrinsic excitation and response relationships, which are essentially the equations of non-equilibrium thermodynamic properties.

(1) Thermal conduction constitutive equation (Fourier law)

Fourier proposed the law of heat conduction through experimental observation in 1822, which generally takes the form of:

$$q_{\mathrm{h}} = -\gamma \nabla T \tag{7.11}$$

Among them, the negative sign indicates that the direction of heat transfer is opposite to the direction of temperature increase, heat flux density q_{h}, that is, the heat flow rate per unit area per unit time; thermal conductivity γ; temperature gradient $\bigtriangledown T$.

(2) Water motion equation (Darcy law)

Darcy proposed the seepage equation, also known as Darcy's law, in 1856 through experimental research on saturated sand. It is generally expressed as:

$$v = -\frac{K}{\mu_{\mathrm{f}}}(\nabla P - \rho_{\mathrm{w}} g) \tag{7.12}$$

Among them, v is the velocity of the fluid, also known as Darcy velocity;Absolute permeability ($K = {}^{k}/_{\rho_{\mathrm{w}} g}$) Gravitational acceleration g.

Since the temperature gradient also has a certain impact on seepage, this phenomenon is commonly referred to as the Soret-like effect. The influence of temperature on water flow movement is typically expressed using empirical equations.

$$q_t = -D_T \nabla T \tag{7.13}$$

Water flow caused by temperature changes q_t, The diffusion coefficient of water flow under temperature difference D_T. This includes the effects of the thermal expansion coefficients of the soil skeleton and pore water, as well as the physicochemical change coefficients.

So, the generalized Darcy's law considering the Soret-like effect can be expressed as:

$$v = -\frac{K}{\mu_f}(\nabla P - \rho_W g) - D_T \nabla T \tag{7.14}$$

According to the Dupuit-Forchheimer relationship:

$$v = \phi v_r \tag{7.15}$$

When considering that the soil skeleton also flows:

$$v_r = \frac{v}{\phi} = v_w - v_s \tag{7.16}$$

Actual flow velocity of fluid v_w; The movement speed of the soil skeleton v_s; The actual average flow velocity of water relative to soil particles v_r.

(3) Thermoelastic constitutive equation

According to the theory of linear thermal stress, the displacements and strains due to elastic deformation caused by external forces can be algebraically superimposed with those due to thermal deformation caused by temperature changes, and the thermal strains resulting from temperature changes are independent of other mechanical stresses. Thus, the strain equation can be expressed as:

$$\varepsilon = \frac{1}{2G}\sigma' - \frac{\gamma}{2G(1+\gamma)}(tr\sigma') + \beta_{Tb}(T - T_0) - \frac{1}{3K_m}(P - P_0) \tag{7.17}$$

In the above equation, the first two terms on the right side represent the strains caused by stress, the third term represents the thermal strain, and the fourth term represents the strain induced by pressure changes. Thus, the total strain consists of these three types of strains combined algebraically.

According to the theory of thermoelasticity, the strain can be expressed as follows:

$$\varepsilon_{ij} = \varepsilon_{ij}^e + \varepsilon_{ij}^T \tag{7.18}$$

Elastic strain of soil skeleton ε_{ij}^e and Thermal strain of soil skeleton ε_{ij}^T. Elastic strain includes strains caused by pressure and those induced by stress. For thermo-hydraulic-mechanical coupling problems, temperature changes will produce certain strains. Under linear assumptions, thermal strain can be expressed as:

$$\varepsilon_{ij}^T = \beta_s(T - T_0) \tag{7.19}$$

For the constitutive equation relating stress, strain, and temperature in homogeneous isotropic materials, it can be expressed as follows in the context of small deformations in thermoelastic problems.

$$\sigma'_{ij} = \lambda \varepsilon_v \delta_{ij} + 2G\varepsilon_{ij} - \beta(T - T_0)\delta_{ij} \tag{7.20}$$

Lame's constants λ; Shear modulus G; Thermal stress coefficient ($\beta = (2G + 3\lambda)\beta_s$); Kronecker Symbol δ_{ij}, when $i = j$, $\delta_{ij} = 1$, when $i = j$, $\delta_{ij} = 0$.

In general, the material properties of the rock skeleton, such as the thermal expansion coefficient, elastic modulus, and Poisson's ratio, vary with temperature. Therefore, it is necessary to consider the relationship between material property coefficients and temperature in the constitutive equations. This results in the relationships among stress, strain, and temperature, as well as the heat conduction equations, being classified as non-linear.

When the temperature is T_0, the pressure is $P = 0$, and the strain is 0, the physical property coefficients are set to λ_0, G_0, $\beta 0$, $\Delta T = T - T_0$. If we consider λ_1, G_1, $\beta 1$ as the physical property coefficients corresponding to the unit relative temperature difference ($\Delta T = 1$), then at a temperature of T, λ, G, β can be expressed as:

$$\lambda = \lambda_0 + \lambda_1 \Delta T = \lambda(T) \tag{7.21}$$

$$G = G_0 + G_1 \Delta T = G(T) \tag{7.22}$$

$$\beta = \beta_0 + \beta_1 \Delta T = \beta(T) \tag{7.23}$$

When considering the effect of temperature on physical property coefficients, the coefficients vary linearly with temperature. Therefore, Eq. (7.20) can be rewritten as:

$$\sigma'_{ij} = \lambda(T)\varepsilon_v \delta_{ij} + 2G(T)\varepsilon_{ij} - \beta(T)(T - T_0)\delta_{ij} \tag{7.24}$$

This constitutive equation can also be expressed in another form, σ^0_{ij} representing the linear part of the above equation as follow:

$$\sigma^0_{ij} = \lambda_0 \varepsilon_v \delta_{ij} + 2G_0 \varepsilon_{ij} - \beta_0(T - T_0)\delta_{ij} \tag{7.25}$$

Nonlinear part $\sigma^I_{ij}\Delta T$:

$$\begin{aligned} \sigma^I_{ij}\Delta T &= \lambda_1 \Delta T \varepsilon_v \delta_{ij} + 2G_1 \Delta T \varepsilon_{ij} - \beta_1 \Delta T (T - T_0)\delta_{ij} \\ &= \left(\lambda_1 \varepsilon_v \delta_{ij} + 2G_1 \varepsilon_{ij} - \beta_1 (T - T_0)\delta_{ij}\right)\Delta T \end{aligned} \tag{7.26}$$

So, there is

$$\sigma_{ij}^{I} = \lambda_1 \varepsilon_v \delta_{ij} + 2G_1 \varepsilon_{ij} - \beta_1 (T - T_0) \delta_{ij} \tag{7.27}$$

So, there is

$$\sigma'_{ij} = \sigma_{ij}^{0} + \sigma_{ij}^{I} \Delta T \tag{7.28}$$

7.3 Supplementary Equations

(1) Geometrical equation

Geometric equations are derived from the continuity of deformation and reflect the relationship between soil skeleton displacement and strain. According to the small deformation assumption and Cauchy equation, the geometric equation is:

$$\varepsilon_{ij} = \frac{1}{2}\left(u_{i,j} + u_{j,i}\right) \tag{7.29}$$

Displacement component u_i.

(2) Effective stress principle

Terzaghi (1923) proposed the concept and principle of effective stress in the study of soil consolidation theory. He believed that the total stress on any plane in a saturated soil is equal to the sum of effective stress and pore water pressure, that is:

$$\sigma = \sigma' + P \tag{7.30}$$

Expanding to the three-dimensional scenario is:

$$\sigma_{ij} = \sigma'_{ij} + P\delta_{ij} \tag{7.31}$$

Some researchers have made revisions based on this, such as:

$$\sigma_{ij} = \sigma'_{ij} + (\alpha - \phi\alpha + \phi)P\delta_{ij} \tag{7.32}$$

Among them, α is the coefficient of pore pressure transmission as particle stress, which is related to the area of pore pressure acting on the particle, $\alpha \in [0, 1]$.

7.4 The Derivation of the Governing Equations

Based on the aforementioned fundamental equations and material property equations, the coupled governing equations for the temperature field, seepage field, and deformation field in the geotechnical medium can be derived as illustrated in Fig. 7.1.

7.4.1 *Governing Equations for the Deformation Field*

The linear momentum balance equation can be derived based on the principle of linear momentum balance.

$$\sigma_{ij,j} + \rho f_i = 0 \tag{7.33}$$

Components of the total stress tensor σij; Equivalent density of the soil medium (soil skeleton and pore water) $\rho = (1 - \phi)\rho_s + \phi\rho_w$; Components of the body force f_i, kg/m^3, when considering only the effect of gravity $f_i = g_i = (0, 0, g)^T$; Gravitational acceleration g, m/s^2.

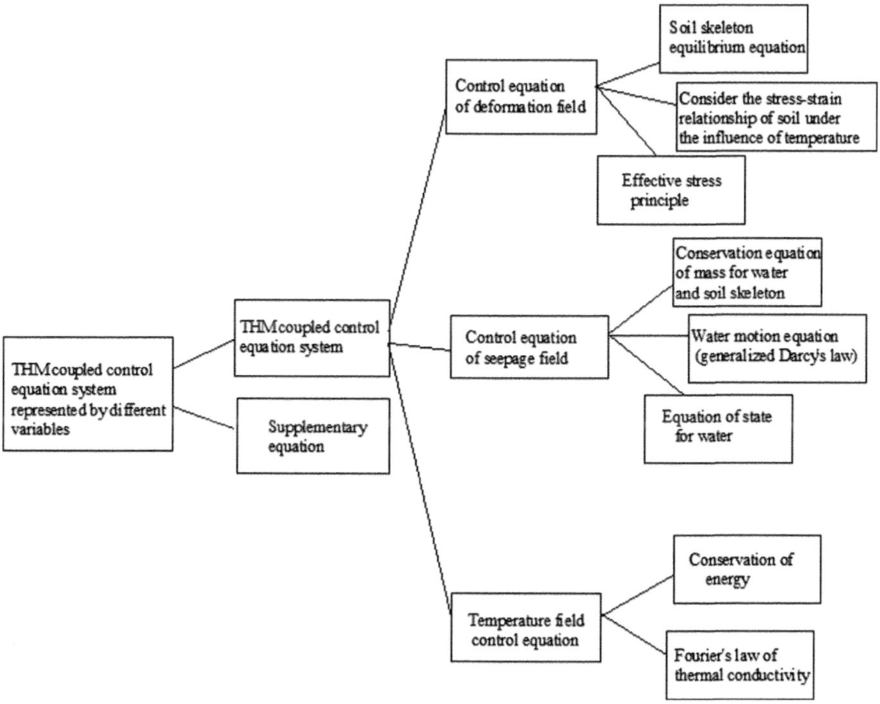

Fig. 7.1 The basic control equations of THM coupling and its derivation process

By substituting (7.32) into (7.33), the static equilibrium equation expressed in terms of effective stress can be obtained.

$$\left[\sigma'_{ij} + (\alpha - \phi\alpha + \phi)P\delta_{ij}\right]_{,j} + \rho f_i = 0 \tag{7.34}$$

By substituting (7.24) into (7.34), the isotropic nonlinear thermoelastic equilibrium equations can be obtained.

$$\begin{aligned}&[\lambda(T)\varepsilon_v\delta_{ij} + 2G(T)\varepsilon_{ij} - \beta(T)(T - T_0)\delta_{ij}\\&+ (\alpha - \phi\alpha + \phi)P\delta_{ij}]_{,j} + \rho f_i = 0\end{aligned} \tag{7.35}$$

The above equation represents the governing equations of the deformation field in an isotropic nonlinear thermoelastic soil medium under the combined effects of the temperature field and seepage field. The last two terms within the brackets on the left side of the equation reflect the influences of thermal stress and pore water pressure on the static equilibrium equation of the soil, fully illustrating and describing the coupling effects of the temperature field and seepage field on the deformation field. By further transforming Eq. (7.35), we can obtain:

$$\begin{aligned}&\left[\lambda u_{k,k} + G\left(u_{i,j} + u_{j,i}\right) - \beta(T - T_0)\delta_{ij} + (\alpha - \phi\alpha + \phi)P\delta_{ij}\right]_{,j}\\&+ [(1 - \phi)\rho_s + \phi\rho_w]f_i = 0\end{aligned} \tag{7.36}$$

$$\begin{aligned}&\lambda_{,j}u_{k,k} + (\lambda + G)\varepsilon_{v,j} + G_{,j}\left(u_{i,j} + u_{j,i}\right) + G\nabla^2 u_i\\&- \beta_{,j}(T - T_0)\delta_{ij} - \beta T_{,j}\delta_{ij} + \left(\alpha - \phi_{,j}\alpha + \phi_{,j}\right)P\delta_{ij}\\&+ (\alpha - \phi\alpha + \phi)P_{,j}\delta_{ij} + [(1 - \phi)\rho_s + \phi\rho_w]f_i = 0\end{aligned} \tag{7.37}$$

The above equation is obtained by finding the derivatives of time t in the three directions of *x*, *y*, *z*:

$$\begin{aligned}&\lambda_1\frac{\partial^2 T}{\partial x\partial t}\varepsilon_v + \lambda_1\frac{\partial T}{\partial x}\cdot\frac{\partial\varepsilon_v}{\partial t} + (\lambda_1 + G_1)\frac{\partial T}{\partial t}\cdot\frac{\partial\varepsilon_v}{\partial x} + (\lambda + G)\frac{\partial^2\varepsilon_v}{\partial x\partial t}\\&+ G_1\frac{\partial T}{\partial t}\nabla^2 u_x + G\frac{\partial\left(\nabla^2 u_x\right)}{\partial t} - (2\beta_1\Delta T + 2\beta_1 + \beta_0)\frac{\partial^2 T}{\partial x\partial t}\\&+ (1 - \alpha)\phi_0 P\left(c_\phi\frac{\partial^2 P}{\partial x\partial t} + \beta_\phi\frac{\partial^2 T}{\partial x\partial t}\right) + \left[\alpha + \phi_0(1 - \alpha)\left(c_\phi\frac{\partial P}{\partial x} + \beta_\phi\frac{\partial T}{\partial x}\right)\right]\frac{\partial P}{\partial t}\\&+ \phi_0(1 - \alpha)\left(c_\phi\frac{\partial P}{\partial t} + \beta_\phi\frac{\partial T}{\partial t}\right)\cdot\frac{\partial P}{\partial x} + (\alpha - \phi\alpha + \phi)\frac{\partial^2 P}{\partial x\partial t} = 0\end{aligned} \tag{7.38}$$

$$\begin{aligned}
&\lambda_1 \frac{\partial^2 T}{\partial y \partial t}\varepsilon_{\rm v} + \lambda_1 \frac{\partial T}{\partial y} \cdot \frac{\partial \varepsilon_{\rm v}}{\partial t} + (\lambda_1 + G_1)\frac{\partial T}{\partial t} \cdot \frac{\partial \varepsilon_{\rm v}}{\partial y} + (\lambda + G)\frac{\partial^2 \varepsilon_{\rm v}}{\partial y \partial t} \\
&+ G_1 \frac{\partial T}{\partial t}\nabla^2 u_y + G\frac{\partial\left(\nabla^2 u_y\right)}{\partial t} - (2\beta_1 \Delta T + 2\beta_1 + \beta_0)\frac{\partial^2 T}{\partial y \partial t} \\
&+ (1-\alpha)\phi_0 P\left(c_\phi \frac{\partial^2 P}{\partial y \partial t} + \beta_\phi \frac{\partial^2 T}{\partial y \partial t}\right) + \left[\alpha + \phi_0(1-\alpha)\left(c_\phi \frac{\partial P}{\partial y} + \beta_\phi \frac{\partial T}{\partial y}\right)\right]\frac{\partial P}{\partial t} \\
&+ \phi_0(1-\alpha)\left(c_\phi \frac{\partial P}{\partial t} + \beta_\phi \frac{\partial T}{\partial t}\right) \cdot \frac{\partial P}{\partial y} + (\alpha - \phi\alpha + \phi)\frac{\partial^2 P}{\partial y \partial t} = 0
\end{aligned} \tag{7.39}$$

$$\begin{aligned}
&\lambda_1 \frac{\partial^2 T}{\partial z \partial t}\varepsilon_{\rm v} + \lambda_1 \frac{\partial T}{\partial z} \cdot \frac{\partial \varepsilon_{\rm v}}{\partial t} + (\lambda_1 + G_1)\frac{\partial T}{\partial t} \cdot \frac{\partial \varepsilon_{\rm v}}{\partial z} + (\lambda + G)\frac{\partial^2 \varepsilon_{\rm v}}{\partial z \partial t} \\
&+ G_1 \frac{\partial T}{\partial t}\nabla^2 u_z + G\frac{\partial\left(\nabla^2 u_z\right)}{\partial t} - (2\beta_1 \Delta T + 2\beta_1 + \beta_0)\frac{\partial^2 T}{\partial z \partial t} \\
&+ (1-\alpha)\phi_0 P\left(c_\phi \frac{\partial^2 P}{\partial z \partial t} + \beta_\phi \frac{\partial^2 T}{\partial z \partial t}\right) + \left[\alpha + \phi_0(1-\alpha)\left(c_\phi \frac{\partial P}{\partial z} + \beta_\phi \frac{\partial T}{\partial z}\right)\right]\frac{\partial P}{\partial t} \\
&+ \phi_0(1-\alpha)\left(c_\phi \frac{\partial P}{\partial t} + \beta_\phi \frac{\partial T}{\partial t}\right) \cdot \frac{\partial P}{\partial z} + (\alpha - \phi\alpha + \phi)\frac{\partial^2 P}{\partial z \partial t} \\
&+ \phi_0 g(\rho_{\rm w} - \rho_{\rm s})\left(c_\phi \frac{\partial P}{\partial t} + \beta_\phi \frac{\partial T}{\partial t}\right) + \phi\rho_{w0} g\left(c_{\rm w}\frac{\partial P}{\partial t} + \beta_{\rm w}\frac{\partial T}{\partial t}\right) - \rho_{\rm s} g\frac{K_{\rm b}}{K_{\rm s}}\frac{\partial \varepsilon_{\rm v}}{\partial t} \\
&+ \frac{\rho_{\rm s}}{K_{\rm s}} g\left[(1-\phi) - \frac{K_{\rm b}}{K_{\rm s}}\right]\frac{\partial P}{\partial t} - 3g\rho_{\rm s}\left[(1-\phi)\beta_{\rm s} - \frac{K_{\rm b}}{K_{\rm s}}\beta_{\rm b}\right]\frac{\partial T}{\partial t} = 0
\end{aligned} \tag{7.40}$$

7.4.2 Governing Equations for the Seepage Field

The continuity equation is essentially a mathematical expression of the law of mass conservation. Based on the law of mass conservation, the continuity equations for water and the soil skeleton can be obtained separately. In seepage mechanics, the characteristics of pressure are of utmost concern. Assuming a saturated porous elastic medium and considering mass sources, the relationship between pressure and density is derived, leading to the commonly used form of the continuity equation:

$$\frac{\partial(\rho\phi)}{\partial t} + \nabla(\rho V) = \rho Q \tag{7.41}$$

The continuity equation for water can be expressed as:

$$\frac{\partial(\rho_{\rm w}\phi)}{\partial t} + \nabla(\rho_{\rm w}\phi v_{\rm w}) = \rho_{\rm w} Q_{\rm w} \tag{7.42}$$

Therefore $v_{\mathrm{w}} = v_{\mathrm{r}} + v_{\mathrm{s}}$ it follows that:

$$\frac{\partial(\rho_{\mathrm{w}}\phi)}{\partial t} + \nabla(\rho_{\mathrm{w}}\phi v_{\mathrm{r}}) + \nabla(\rho_{\mathrm{w}}\phi v_{\mathrm{s}}) = \rho_{\mathrm{w}}Q_{\mathrm{w}} \tag{7.43}$$

Partial expansion can reveal

$$\frac{\partial(\rho_{\mathrm{w}}\phi)}{\partial t} + \nabla(\rho_{\mathrm{w}}\phi v_{\mathrm{r}}) + \rho_{\mathrm{w}}\phi\nabla v_{\mathrm{s}} + v_{\mathrm{s}}\nabla(\rho_{\mathrm{w}}\phi) = \rho_{\mathrm{w}}Q_{\mathrm{w}} \tag{7.44}$$

Due to the small size of item $v_{\mathrm{s}}\nabla(\rho_{\mathrm{w}}\phi)$, it can be ignored:

$$\frac{\partial(\rho_{\mathrm{w}}\phi)}{\partial t} + \nabla(\rho_{\mathrm{w}}V) + \rho_{\mathrm{w}}\phi\nabla v_{\mathrm{s}} = \rho_{\mathrm{w}}Q_{\mathrm{w}} \tag{7.45}$$

Among them, Q_{w} is the source sink strength within the unit, which refers to the volume of fluid generated or engulfed by a unit volume per unit time.

The continuity equation of the soil skeleton is:

$$\frac{\partial[\rho_{\mathrm{s}}(1-\phi)]}{\partial t} + \nabla[\rho_{\mathrm{s}}(1-\phi)v_{\mathrm{s}}] = 0 \tag{7.46}$$

It can also be ignored $v_{\mathrm{s}}\nabla[\rho_{\mathrm{s}}(1-\phi)]$ Multiply each item again $\rho_{\mathrm{w}}/\rho_{\mathrm{s}}$ available

$$\frac{(1-\phi)\rho_{\mathrm{w}}}{\rho_{\mathrm{s}}}\frac{\partial\rho_{\mathrm{s}}}{\partial t} - \rho_{\mathrm{w}}\frac{\partial\phi}{\partial t} + \rho_{\mathrm{w}}(1-\phi)\nabla v_{\mathrm{s}} = 0 \tag{7.47}$$

And because:

$$\nabla v_{\mathrm{s}} = \frac{\partial}{\partial t}(\nabla u_{\mathrm{s}}) = \frac{\partial}{\partial t}\left(\delta_{ij}\varepsilon_{ij}\right) = \frac{\partial\varepsilon_{\mathrm{v}}}{\partial t} = \frac{\partial}{\partial t}\left(u_{i,i}\right) \tag{7.48}$$

Among them, u_{s} is the displacement vector of the soil skeleton.

By adding Eqs. (7.45) and (7.47) and expanding them, the overall continuity equation can be obtained:

$$\phi\frac{\partial\rho_{\mathrm{w}}}{\partial t} - \nabla\left[\frac{\rho_{\mathrm{w}}K}{\mu_{\mathrm{f}}}(\nabla P - \rho_{\mathrm{w}}g)\right] + \frac{(1-\phi)\rho_{\mathrm{w}}}{\rho_{\mathrm{s}}}\frac{\partial\rho_{\mathrm{s}}}{\partial t} + \rho_{\mathrm{w}}\frac{\partial\varepsilon_{\mathrm{v}}}{\partial t} = \rho_{\mathrm{w}}Q_{\mathrm{w}} \tag{7.49}$$

It can be inferred from (7.5) that:

$$\frac{\partial\rho_{\mathrm{w}}}{\partial t} = \rho_{w0}\left(c_{\mathrm{w}}\frac{\partial P}{\partial t} - \beta_{\mathrm{w}}\frac{\partial T}{\partial t}\right) \tag{7.50}$$

Substitute $\frac{\partial \rho_w}{\partial t}$, $\frac{\partial \rho_s}{\partial t}$ into Eq. (7.49), we obtain:

$$\rho_w\left(1-\frac{K_b}{K_s}\right)\frac{\partial \varepsilon_v}{\partial t}+\left[\rho_{w0}\phi c_w+\frac{\rho_w}{K_s}\left(1-\frac{K_b}{K_s}-\phi\right)\right]\frac{\partial P}{\partial t}$$
$$-\left\{3\rho_w\left[(1-\phi)\beta_s-\frac{K_b}{K_s}\beta_b\right]+\rho_{w0}\phi\beta_w\right\}\frac{\partial T}{\partial t}$$
$$-\nabla\left[\frac{\rho_w K}{\mu_f}(\nabla P-\rho_w g)+\rho_w D_T\nabla T\right]=\rho_w Q_w \tag{7.51}$$

Equation (7.51) represents the governing equation of the seepage field under the combined influence of the deformation field and the temperature field. The first term on the left side reflects the coupling between the deformation field and the seepage field, indicating changes in the seepage field caused by soil skeleton deformation. The parts containing the temperature variable T in the third and last terms on the left side reflect the coupling between the temperature field and the seepage field, indicating changes in the seepage flow pattern due to temperature variations. Therefore, this equation fully captures the coupling effect of the temperature field and deformation field on the seepage field.

7.4.3 Temperature Field Control Equation

The heat transfer time between soil particles and water in the soil is very brief, assuming that the soil particles and pore water are in a local thermal equilibrium state. However, during the derivation of the governing equation, the assumption of no local thermal equilibrium is made, and the thermal balance for the soil skeleton and water is derived separately. An infinitesimal parallelepiped within the soil is taken as the equilibrium unit, with edge lengths of dx, dy, dz, as shown in Fig. 7.2.

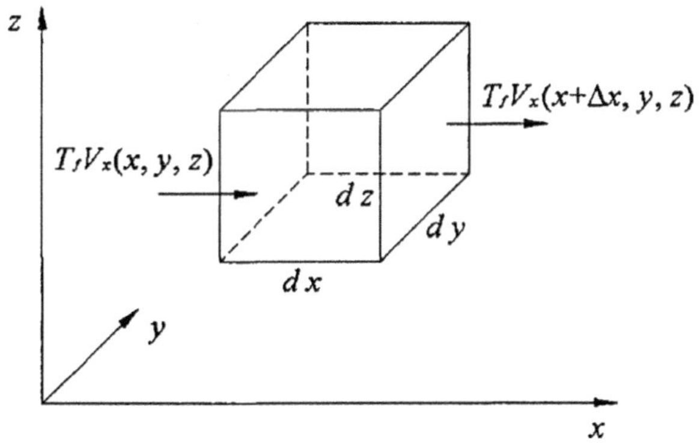

Fig. 7.2 Heat migration unit cell schematic diagram

Porosity of soil is ϕ. The volume of the soil skeleton within the unit is $(1-\phi)\mathrm{d}x\mathrm{d}y\mathrm{d}z$. The volume of pore water is $\phi\mathrm{d}x\mathrm{d}y\mathrm{d}z$. Below, we will discuss the thermal equilibrium of pore water in the soil skeleton within the unit.

(1) Thermal equilibrium equation of pore water

The factors causing changes in the temperature of pore water within the unit during time dt mainly include: heat conduction of the water, heat convection caused by the movement of pore water, and heat exchange between the soil skeleton and pore water.

1) Thermal conduction effect

On a macroscopic scale, the heat flux density q (the amount of heat transferred per unit area perpendicular to the direction of heat flow) is proportional to the temperature gradient, which is referred to as Fourier's law of heat conduction mentioned above. The difference in heat entering and leaving the unit's pore water due to heat conduction along the x-axis during time dt is given by:

$$\begin{aligned}&\left[-\gamma_{wx}\frac{\partial T}{\partial x}-\left(-\gamma_{wx}\frac{\partial T}{\partial x}+\frac{\partial}{\partial x}\left(-\gamma_{wx}\frac{\partial T}{\partial x}\right)\mathrm{d}x\right)\right]\phi\mathrm{d}y\mathrm{d}z\mathrm{d}t\\&\quad=\frac{\partial}{\partial x}\left(-\gamma_{wx}\frac{\partial T}{\partial x}\right)\phi\mathrm{d}xy\mathrm{d}z\mathrm{d}t,\end{aligned}\tag{7.52}$$

where λ_w is thermal conductivity of water. Similarly, the heat difference along the y and z directions is:

$$\frac{\partial}{\partial y}\left(-\gamma_{wy}\frac{\partial T}{\partial y}\right)\phi\mathrm{d}x\mathrm{d}y\mathrm{d}z\mathrm{d}t,\ \frac{\partial}{\partial z}\left(-\gamma_{wz}\frac{\partial T}{\partial z}\right)\phi\mathrm{d}x\mathrm{d}y\mathrm{d}z\mathrm{d}t$$

Therefore, the difference in heat of the pore water entering and leaving the unit during time dt is given by:

$$\nabla(\gamma_\mathrm{w}\nabla T)\phi\mathrm{d}x\mathrm{d}y\mathrm{d}z\mathrm{d}t$$

2) The heat convection effect

Heat convection refers to the phenomenon where heat is transferred from one location to another through a moving medium. Convection occurs only in fluids, and because molecules in a fluid undergo random thermal motion, convection is always accompanied by heat conduction. The difference in heat of the pore water entering and leaving the unit along the x-axis due to heat convection during time is dt given by:

$$\begin{aligned}&\left[\rho_\mathrm{w}C_\mathrm{f}Tv_{rx}(x,y,z)-\rho_\mathrm{w}C_\mathrm{f}Tv_{rx}(x+\mathrm{d}x,y,z)\right]\phi y\mathrm{d}z\mathrm{d}t\\&\quad=-\rho_\mathrm{w}C_\mathrm{f}\frac{\partial(Tv_{rz})}{\partial x}\phi\mathrm{d}x\mathrm{d}y\mathrm{d}z\mathrm{d}t,\end{aligned}\tag{7.53}$$

where C_f is the specific heat capacity of water, V_r is the seepage velocity of water relative to the soil skeleton. Similarly, due to convection, the difference in heat along the y and z directions is given by:

$$-\rho_w C_f \frac{\partial (Tv_{rz})}{\partial y} \phi \mathrm{d}x\mathrm{d}y\mathrm{d}z\mathrm{d}t, -\rho_w C_f \frac{\partial (Tv_{rz})}{\partial z} \phi \mathrm{d}x\mathrm{d}y\mathrm{d}z\mathrm{d}t$$

Therefore, the difference in heat of the pore water entering and leaving the unit due to heat convection during time dt is given by:

$$\begin{aligned} &-\rho_w C_f \nabla (Tv_r) \phi \mathrm{d}x\mathrm{d}y\mathrm{d}z\mathrm{d}t = \rho_w C_f \nabla \\ &\left(T \frac{k}{\mu_f} (\nabla P - \rho_W g) + D_T \nabla T \right) \mathrm{d}x\mathrm{d}y\mathrm{d}z\mathrm{d}t \end{aligned} \quad (7.54)$$

The total heat changes in pore water due to the various effects mentioned above will cause a temperature change dt over time. The amount of heat required to cause a temperature change dT in pore water with a volume $\phi \mathrm{d}x\mathrm{d}y\mathrm{d}z$ is given by:

$$\rho_w C_f \phi \mathrm{d}x\mathrm{d}y\mathrm{d}z\mathrm{d}T$$

According to the law of conservation of energy, the heat change caused by the aforementioned heat conduction and convection effects should be equal to $\rho_w C_f \phi \mathrm{d}x\mathrm{d}y\mathrm{d}z\mathrm{d}T$. Thus, the thermal equilibrium equation for the pore water can be expressed as:

According to the law of conservation of energy, the heat change caused by the aforementioned heat conduction and convection effects should be equal to the heat added or removed. Therefore, the thermal equilibrium equation for the pore water can be expressed as:

$$\rho_w C_f \phi \frac{\partial T}{\partial t} = \nabla (\phi \gamma_w \nabla T) + \rho_w C_f \nabla \left(T \frac{k}{\mu_f} (\nabla P - \rho_W g) + D_T \nabla T \right) \quad (7.55)$$

(2) The thermal equilibrium equation for the soil skeleton

The primary factor causing temperature changes in the soil skeleton is heat conduction within the soil skeleton. Based on the derivation of the pore water thermal equilibrium equation, the thermal equilibrium equation for the soil skeleton can be expressed as:

$$\rho_s C_s (1 - \phi) \frac{\partial T}{\partial t} = \nabla \left[(1 - \phi) \gamma_s \nabla T \right] \quad (7.56)$$

(3) The heat source/sink term

The heat source/sink term primarily includes the heat generation rate within the soil itself as well as other various additional heat sources/sinks.

1) Heat generation rate

The heat generation rate of the soil itself is generally attributed to heat changes caused by internal friction during deformation. The heat generation rate function in general can be expressed as follows:

$$\frac{\rho T \lambda_{\mathrm{e}}}{\varphi} = -\rho \int_{-\infty}^{\zeta} \int_{-\infty}^{\zeta} \frac{\partial J_{ijkl}}{\partial \zeta}(\zeta - \zeta_1, \zeta - \zeta_2) \frac{\partial \varepsilon_{ij}^*}{\partial \zeta_1} \frac{\partial \varepsilon_{kl}^*}{\partial \zeta_2} \mathrm{d}\zeta_1 \mathrm{d}\zeta_2 \geq 0, \tag{7.57}$$

where λ_{e} is the entropy increase caused by internal friction during deformation; J_{ijkl} is a continuous kernel function; φ is the temperature equivalence coefficient; variable $\zeta = T\varphi$, $\varepsilon_{ij}^* = \varepsilon_{ij} - \theta\alpha_{ij}$, α_{ij} Anisotropic coefficient of thermal expansion, θ is temperature increment, and it is believed here that it is unrelated to the history of deformation.

2) Thermoelastic coupling term

Theoretical and experimental studies have shown that the deformation of the soil skeleton affects the distribution of the temperature field. The thermoelastic coupling term is expressed as:

$$Q_{\mathrm{s}} = -(1 - \phi) T_* \beta \frac{\partial \varepsilon_{\mathrm{v}}}{\partial t} \tag{7.58}$$

where T_* is absolute temperature.

3) Viscous dissipation

Viscous dissipation is the phenomenon where mechanical energy is converted into thermal energy due to viscous friction forces during fluid flow. This process leads to changes in the fluid's temperature, viscosity, and heat transfer behavior, thereby affecting the flow characteristics of the fluid. The effect of viscous dissipation is equivalent to the heat generated by an internal heat source being transferred along the flow direction of the fluid. The amount of heat generated per unit volume due to viscous dissipation is referred to as the viscous dissipation heat rate.

$$Q_{\mathrm{f}} = \frac{\mu_{\mathrm{f}}}{k} v \cdot v + \frac{\phi F \rho_{\mathrm{w}}}{\sqrt{k}} v|v|v, \tag{7.59}$$

where v is average fluid velocity, F is inertia constant, $F = b/\sqrt{a\phi^{2/3}}$, a, b are empirical constants.

4) The heat generated by the radioactive decay of isotopes

In nuclear waste management, some high-level radioactive waste generates heat due to radioactive decay. We assume that the heat generated by the decay of nuclear waste decreases exponentially over time.

$$Q_{\mathrm{m}} = Q_0 \mathrm{e}^{-\eta t} \tag{7.60}$$

where Q_0 is the initial intensity of the heat source, usually taken as 0.05–0.5 W/m^3; η is attenuation index, usually taken as 0.02/yr.

5) The energy change caused by the reversible deformation of pore water

Assuming that the pore water behaves as an isotropic elastic material, the energy change due to reversible deformation can be expressed as:

$$Q_n = -T_* \frac{\beta_w}{c_w} \nabla \cdot v \tag{7.61}$$

where β_w is volume thermal expansion coefficient of water, c_w is volume compression coefficient of water.

(4) Governing equation

According to the principle of energy conservation, by summing the thermal equilibrium equations for the pore water and soil skeleton, along with the various source terms, the governing equation for the temperature field can be derived as:

$$\begin{aligned}&\rho_w C_f \phi \frac{\partial T}{\partial t} + \rho_s C_s (1-\phi) \frac{\partial T}{\partial t} = \nabla(\phi \gamma_w \nabla T) + \nabla\big[(1-\phi)\gamma_s \nabla T\big] + \\ &\rho_w C_f \nabla\left(T \frac{K}{\mu_f}(\nabla P - \rho_w g) + D_T \nabla T\right) + \frac{\rho T \lambda_e}{\varphi} + Q_s + Q_f + Q_m + Q_n\end{aligned} \tag{7.62}$$

Assuming that the soil skeleton and pore water are in a state of local thermal equilibrium, the above equation can be simplified to:

$$\begin{aligned}\rho C \frac{\partial T}{\partial t} &= \nabla(\gamma \nabla T) + \rho_w C_f \nabla\left[T \frac{K}{\mu_f}(\nabla P - \rho_w g) + D_T \nabla T\right] \\ &+ \frac{\rho T \lambda_e}{\varphi} + Q_s + Q_f + Q_m + Q_n,\end{aligned} \tag{7.63}$$

where $\rho = \rho_w \phi + \rho_s (1-\phi)$ is equivalent density of soil mass; $C = C_f \phi + C_s (1-\phi)$ is equivalent specific heat; $\gamma = \phi \gamma_w + (1-\phi)\gamma_s$ is effective thermal conductivity.

The above equation represents the temperature field control equation for the soil under the assumption that the rock matrix and groundwater are in a state of local thermal equilibrium. Equations (7.37), (7.51), and (7.63) form the set of governing equations for the coupled temperature field, seepage field, and deformation field in the case of isotropic thermoelasticity. The primary variables in these equations are temperature, displacement, and pore water pressure.

7.5 Mechanisms of Coupling Among the Three Fields and Model Simplification for Computation

(1) Mechanisms of Coupling (Fig. 7.3)

The coupling mechanisms between the fields shown in Fig. 7.4 can be described as:

1) M → H, Deformation of the soil alters the porosity, which in turn changes the permeability. Additionally, deformation-induced changes in pore water pressure also affect the permeability.
2) H → M, Changes in pore water pressure affect deformation, and the physico-chemical interactions between pore water and the soil also play a role in this process.
3) M → T, Deformation of the soil skeleton leads to changes in the pore structure, which in turn affects the thermal properties of the soil and consequently influences the temperature field. Additionally, internal dissipative processes within the soil contribute to changes in the temperature field.
4) T → M, Thermal stresses and thermal strains induced by temperature affect the stress field and deformation field. Additionally, temperature changes impact the physical and mechanical properties of the soil as well as its thermal characteristics.
5) H → T, The convective heat transfer of the seepage fluid affects the temperature field. Additionally, the work done by fluid compression and viscous dissipation also influences the temperature field.

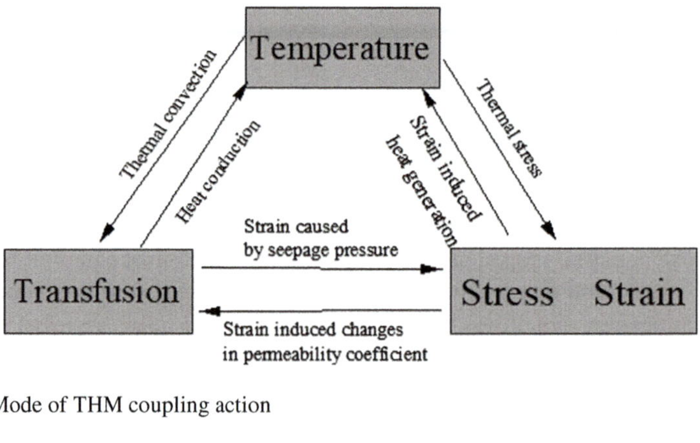

Fig. 7.3 Mode of THM coupling action

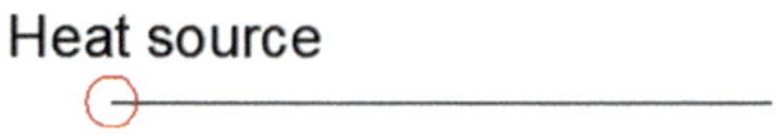

Fig. 7.4 One dimensional point heat source schematic diagram

6) T → H, Changes in pore structure caused by temperature variations lead to alterations in permeability. Temperature changes also affect the physical, mechanical, and thermal properties of the seepage fluid. Additionally, seepage flow induced by temperature gradients impacts the overall seepage behavior.

Based on the different modes of coupling, the mechanisms of coupling among the three fields can be categorized into mechanical coupling and parametric coupling.

Mechanical coupling refers to the coupling between fields through specific mechanical actions or processes. For example: The temperature field couples with the deformation field through thermal deformation induced by temperature changes. The seepage field couples with the temperature field through convective heat transfer in the fluid. The deformation field couples with the seepage field through mechanical interactions affecting the soil's pore structure. Parametric coupling refers to the coupling effects manifested through changes in the physical parameters of the soil due to variations in temperature and pressure. For instance: Temperature affects the density and viscosity of water, demonstrating coupling with the seepage field. Temperature influences the thermal parameters of the soil, and changes in these thermal parameters subsequently affect the distribution of the temperature field.

(2) Simplification of the Coupling Model

The three-field coupling model described above considers variations in medium density and porosity with temperature and pressure, employs nonlinear elastic constitutive equations, and includes thermal source terms such as viscous dissipation, making the model particularly complex to compute. In practical engineering applications of the three-field coupling theory, the model can be simplified based on preliminary analysis of the research subject.

Generally, the interactions between the deformation field and the seepage field are strong, and the impact of the seepage field on the temperature field is significant. The effect of temperature on the thermal properties of the soil is also noteworthy. However, the impact of the deformation field on the temperature field is relatively minor. In applying the three-field coupling model, certain couplings can be neglected, such as the influence of gravitational effects on seepage, seepage induced by temperature gradients, and the effects of soil skeleton deformation and water viscous dissipation on the temperature field.

To simplify the deformation field control equation, where G, λ, β is considered a constant and the $(a - \phi a + \phi)$ in the effective stress is replaced by the α coefficient, it can be expressed as:

$$\nabla\left[\lambda\varepsilon_{\mathrm{v}}\delta_{ij} + 2G\varepsilon_{ij} - \beta(T - T_0)\delta_{ij} + \alpha P\delta_{ij}\right] + \rho f_i = 0 \tag{7.64}$$

The simplified form of the seepage field control equation, ignoring mass sources, gravitational effects, and the influence of temperature gradients on seepage, it can be expressed as:

$$\rho_{\mathrm{w}}\frac{\partial\varepsilon_{\mathrm{v}}}{\partial t} + \rho_{w0}\phi c_{\mathrm{w}}\frac{\partial P}{\partial t} - \rho_{w0}\phi\beta_{\mathrm{w}}\frac{\partial T}{\partial t} - \nabla\left(\frac{\rho_{\mathrm{w}}K}{\mu_{\mathrm{f}}}\nabla P\right) = 0 \tag{7.65}$$

The simplified form of the temperature field control equation, assuming local thermal equilibrium and neglecting the effects of internal heat sources, can be expressed as:

$$\rho C\frac{\partial T}{\partial t} = \nabla(\gamma\nabla T) + \rho_{\mathrm{w}}C_{\mathrm{f}}\nabla\left(T\frac{K}{\mu_{\mathrm{f}}}\nabla P\right) \tag{7.66}$$

The simplified control equations for the three fields, as described above, represent the most commonly used expressions in current practice.

(3) Calculate and solve

If body forces and water source inputs are not considered, and many researchers have pointed out that the coupling effects between the displacement field and the stress field on the temperature field can be neglected, with the densities of water and soil particles, as well as various coefficients, treated as constants within the solution range, the simplified three-field coupling model can be expressed as:

$$G\frac{\partial\left(\nabla^2 u_i\right)}{\partial t} + (G+\lambda)\frac{\partial^2\varepsilon_{\mathrm{v}}}{\partial t\partial i} - (2G+3\lambda)\beta_{\mathrm{s}}\frac{\partial^2 T}{\partial t\partial i} - \frac{\partial^2 p}{\partial t\partial i} = 0 \tag{7.67}$$

$$\frac{K}{\mu_{\mathrm{f}}}\nabla^2 P - \frac{\partial\varepsilon_{\mathrm{v}}}{\partial t} - c_{\mathrm{w}}\phi\frac{\partial P}{\partial t} + \beta_{\mathrm{w}}\phi\frac{\partial T}{\partial t} = 0 \tag{7.68}$$

$$\gamma\nabla^2 T - \rho C\frac{\partial T}{\partial t} = 0 \tag{7.69}$$

For the case of a one-dimensional infinite medium with a constant heat source power, the distributions of temperature, pore pressure, and displacement can be determined using the heat source function method as follows:

$$T(x,t) = \frac{q}{4\pi\gamma}E\left(\frac{x^2}{4\varepsilon t}\right) \tag{7.70}$$

$$p(x,t) = \frac{c\gamma_{\mathrm{w}}R_0 q}{4\pi km(c-\int)}\left[E\left(\frac{x^2}{4ct}\right) - E\left(\frac{x^2}{4\varepsilon t}\right)\right] \tag{7.71}$$

$$u(x,t) = \frac{q}{2\pi kMm(c-\varepsilon)}\int_0^t H_{,x}(x,t)\mathrm{d}\tau \tag{7.72}$$

Among:

$$\begin{aligned} H(x,t) = {} & c^2\gamma_{\mathrm{w}}R_0[\ln x + E(\omega_\alpha)/2] \\ & - \left[c\gamma_{\mathrm{w}}R_0 + \beta_0 k(c-\varepsilon)\right]\left[\ln x + E\left(\omega_\beta\right)/2\right] \end{aligned} \tag{7.73}$$

$$E(x) = \int_{x}^{\infty} \mathrm{e}^{(-\eta)}/\eta \mathrm{d}\eta, \tag{7.74}$$

where q is heat source power, $m = (1 - \phi)\rho_s C_s + \phi\rho_w C_w$, $c = k/\left[\gamma_w(\phi c_w + 1/M)\right]$, $\epsilon = \gamma/m$, $M = \lambda + 2G$, $R_0 = \phi\beta_w - (2G + 3\lambda)\beta_s/M$, $\omega_\alpha = x^2/4c(t - \tau)$, $\omega_\beta = x^2/4\varepsilon(t - \tau)$

Given the coefficients, the evolution of temperature, pore pressure, and displacement around the heat source can be determined through numerical integration. The coefficients for the experimental soil samples are as follows:

Elastic modulus ($E = 6.0 \times 10^5$ Pa), Poisson's ratio ($v = 0.3$), Void ratio ($\phi = 0.4$), Permeability coefficient ($k = 1.2 \times 10^{-8} \mathrm{m}^2\ \mathrm{s}^{-1}$), Density of soil particles ($p_s = 2600$ kg/m^3), Water density ($\rho_w = 2600$ kg/m^3), Thermal conductivity coefficient ($y = 0.5$W/(m °C)), Bulk modulus of soil particles ($k_s = 2 \times 10^{10}$ Pa), The bulk modulus of water ($k_w = 5 \times 10^9$ Pa), Specific heat capacity ($C_s = 4200$ J/(kg °C)), The specific heat of water and the linear expansion coefficient of soil particles ($\beta_s = 1.5 \times 10^{-5}$/°C), The coefficient of linear expansion of water ($\beta_w = 2.0 \times 10^{-4}$/°C), Specific heat of soil particles ($C_s = 800$ J/(kg °C)), Volume compression coefficient of water ($c_w = 5 \times 10^{-10}$ m^2/N), by substituting the above equation, we can obtain:

$$T(x, t) = \frac{q}{6.28} \int_{1.47\times10^6 x^2/t}^{\infty} \mathrm{e}^{-\eta}/\eta \mathrm{d}\eta \tag{7.75}$$

$$p(x, t) = 1.43q\left(\int_{2.58\times10^5 x^2/t}^{\infty} \mathrm{e}^{-\eta}/\eta \mathrm{d}\eta - \int_{1.47\times10^6 x^2/t}^{\infty} \mathrm{e}^{-\eta}/\eta \mathrm{d}\eta\right) \tag{7.76}$$

$$u(x, t) = 6.99 \times 10^{-13} q \int_{0}^{t} \left(4.92 \times \frac{1 - \mathrm{e}^{-2.58\times10^5 x^2/(t-\tau)}}{x} - 3.03 \times \frac{1 - \mathrm{e}^{-1.47\times10^6 x^2/(t-\tau)}}{x}\right) \mathrm{d}\tau \tag{7.77}$$

When both the heating and cooling power are set to 500 W, numerical solutions to the equation above can be obtained using MATLAB software. This reveals the changes over time in temperature, pore pressure, and displacement at different distances from the heat source (0.01, 0.04, 0.07 m), as shown in Fig. 7.5. The heating phase lasts from 0 to 10,000 s, followed by the cooling phase from 10,000 to 20,000 s.

Figure 7.5 illustrates the relationship between temperature, air pressure, and displacement changes over time under the combined effects of three fields, given a constant power output from the heat source within specific distances and time intervals. In practice, cooling is generally natural and the relationship between temperature and time is complex. For computational convenience, it is assumed that the cooling

Fig. 7.5 Temperature, pore water pressure and displacement varying with time

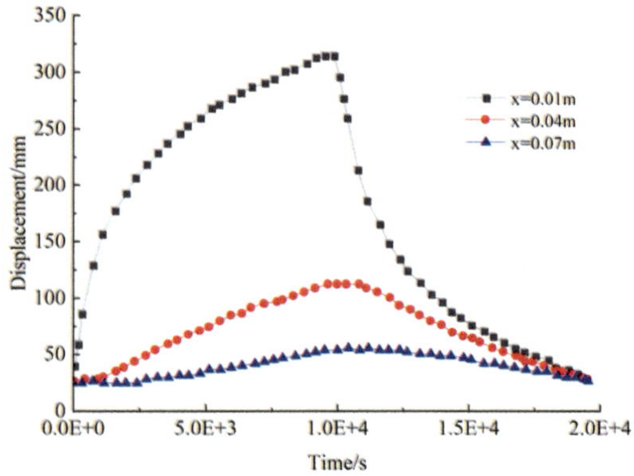

(a) Temperature time history

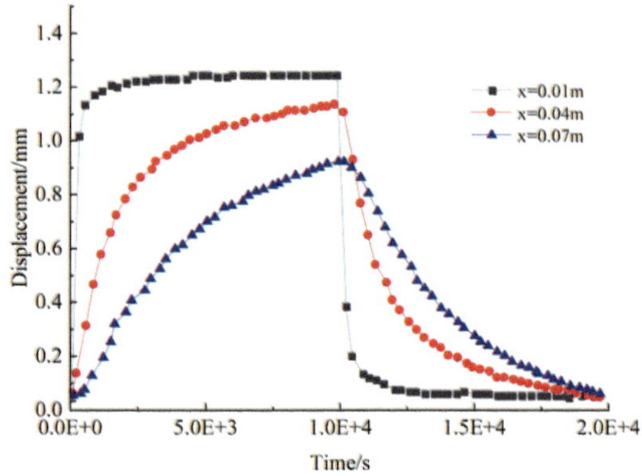

(b) Pore water pressure time history

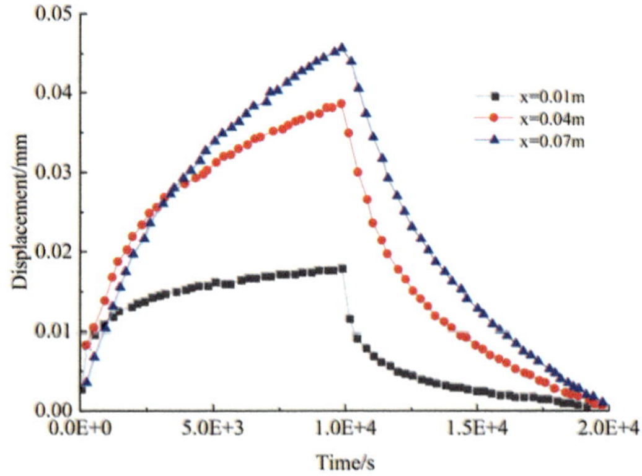

(c) Displacement time history

process also has a defined cooling power, which is the same as the heating power. The observations are as follows:

(1) Temperature Changes: During heating, the farther from the heat source, the slower the rate of change and the lower the maximum temperature reached over time. Initially, the rate of change is rapid, but it gradually decreases.
(2) Pore Pressure Changes: Similar to temperature, during heating, the rate of change and the maximum values decrease with increasing distance from the heat source. The initial rate of change is fast, which then slows down and stabilizes. The farther from the heat source, the longer it takes to stabilize.
(3) Displacement Changes: During heating, the rate of change is initially fast but slows down over time. The closer to the heat source, the faster the rate of change and the shorter the time required to stabilize, corresponding to the changes in pore pressure. At greater distances from the heat source, displacement can stabilize after a sufficiently long period and shows a trend of increasing. During cooling, assuming the power is the same as during heating, the changes mirror those during heating, ultimately returning to the initial values.

7.6 Numerical Analysis of Thermal Consolidation Tests

7.6.1 Nonlinear Elastic Analysis

When the confining pressure is constant (i.e. $\sigma_3\prime$ is constant), the stress changes induced by temperature and the changes in pore water pressure essentially represent a process of loading or unloading for the solid matrix (i.e., a process where the effective stress increases or decreases). Thus, according to linear elastic theory, the volumetric strain in the solid matrix changes linearly with effective stress during loading and unloading. Furthermore, when the effective stress returns to its initial state, the incremental volumetric strain returns to zero.

In fact, the ultimate volumetric strain in thermal consolidation increases with the number of loading–unloading cycles (changes in thermal stress and water pressure). During the loading process, the solid matrix becomes denser, which increases its stiffness matrix. When unloading occurs, the stiffness matrix of the solid matrix is greater than during the loading phase. Similarly, during the unloading process, as the solid matrix becomes looser, its stiffness matrix decreases again. If loading is performed at this point, the stiffness matrix of the solid matrix is smaller than during the previous unloading process. As shown in Fig. 7.6, the elastic modulus E_b of the soil increases gradually during loading (isothermal drained consolidation). During unloading (heating or cooling undrained consolidation), E_b gradually decreases. Since E_b decreases with the number of loading and unloading cycles, this causes the volumetric strain to increase with the number of cycles. That is to say, during the thermal consolidation process, the elastic modulus of the porous medium is influenced by temperature loads and water pressure. In summary, the elastic modulus

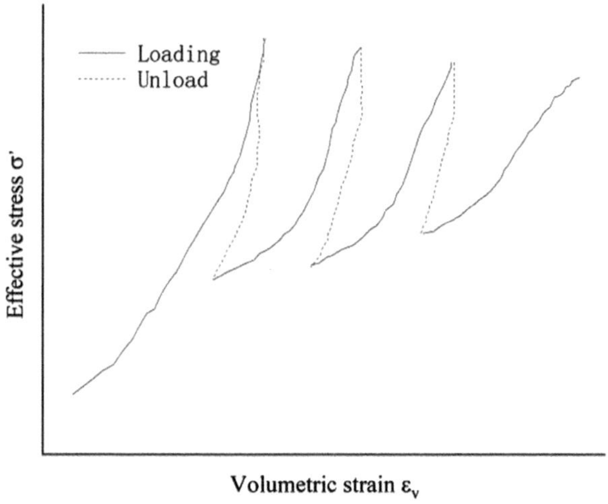

Fig. 7.6 Loading–unloading process in saturated porous media

E_b of the porous medium can be adjusted by the temperature T and the pore water pressure P_w, and it is assumed that the Poisson's ratio υ of the soil remains constant. The simplified model is presented in the following equation:

$$E_b = E_{b,0}\left\{[1-\eta_T(T-T_{N-1})]\prod_{i=2}^{N-1}[1-\eta_T\xi_i(T_i-T_{i-1})] - \frac{\eta_P}{\sigma_3'}\left[\prod_{i=1}^{N}\gamma_i(P_w-P_{w,0})+\sum_{j=1}^{j=i}(1-\gamma_j)(P_{w,j}-P_{w,0})\prod_{j=1}^{N-1}\gamma_{j-1}\right]\right\}, \quad (7.78)$$

where E_b is elastic modulus of porous media, $E_{b,0}$ is initial elastic modulus of porous media, T_i is temperature at the end of the ith drained consolidation, N is number of isothermal drained consolidations; η_T, η_P are correction factors for temperature and pore water pressure, respectively; $\sigma_3\prime$ is consolidation confining pressure; Π is the symbol for repeated multiplication.

7.6.2 Calculation Model and Parameters

Based on the model proposed, numerical simulations of the experiment described in Chap. 5 were conducted. The physical model is shown in Fig. 7.7.

The sample's bottom surface is thermally insulated and has no-flow boundaries, with no vertical displacement allowed; the top surface is a thermal insulation boundary with no vertical displacement allowed; the outer side surfaces have no-flow boundaries, with the temperature set to Tout and pressure set to σ_3' the inner side surfaces also have no-flow boundaries, with the temperature set to Tin and no

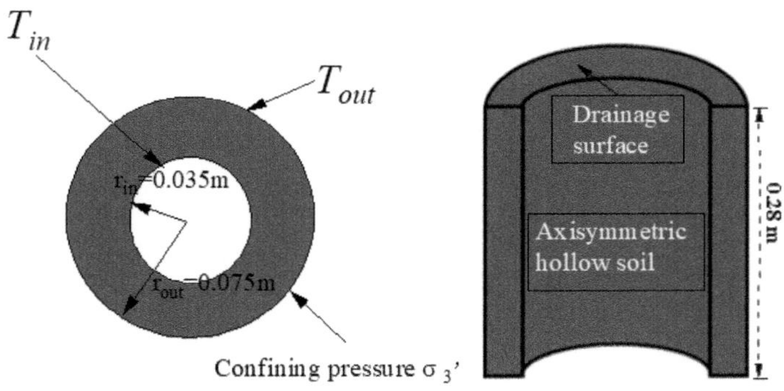

(a) Cross section (b) Longitudinal section

Fig. 7.7 Schematic diagram of the analysis model

vertical displacement allowed. The experiment initially proceeds with isothermal (room temperature) drained consolidation under confining pressure σ_3', with the drainage face being the top surface of the soil body, as shown in Fig. 7.7. Then, the sample undergoes heating without drainage under different temperature loads applied to the inner and outer side surfaces. After the temperatures on the inner and outer surfaces stabilize, another isothermal drained consolidation is conducted. The parameters for the sample are listed in Table 7.1.

The heating paths for the inner and outer boundaries during the experimental process are depicted in Fig. 7.8. In the diagram, the dark gray area represents the isothermal drained consolidation process, while the light gray area corresponds to the undrained heating consolidation process. The duration for the undrained heating consolidation is set to 200 min in the calculations. The isothermal drained consolidation process lasts 20 min, and it is assumed that the temperature–time relationship for the inner and outer side surfaces during the heating process is linear. Under each heating path, calculations are performed for three different confining pressures: $\sigma_3\prime$ = 50 kPa, 100 kPa, and 150 kPa.

Based on the Comsol Multiphysics numerical analysis platform, a program is developed to perform coupled analysis of the described thermal consolidation model, which includes modifying control equations, defining related parameter variables, and specifying the methods for transferring calculation results between different consolidation processes. The seepage equation utilizes the "Darcy's Law" physics

Table 7.1 Model parameters

Parameter values	ρ_s	$E_{b,0}$	E_m	n	γ	k	υ	c_s	β_s	β_{Tb}
	g/cm^3	MPa	MPa		W/(m K)	m^2		J/(kg K)	1/K	1/K
	1.66	0.5	100	0.59	1.39	6.48×10^{-12}	0.4	1334	3.1×10^{-6}	9.3×10^{-6}

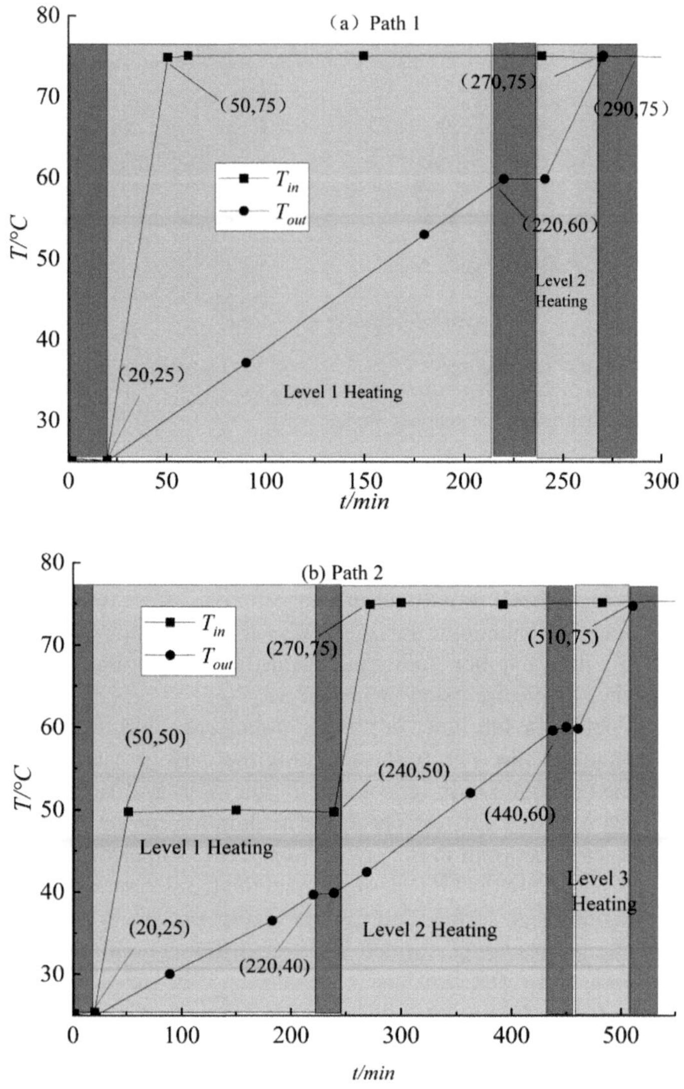

Fig. 7.8 Heating path of porous medium in calculation

interface, the stress balance equation employs the "Solid Mechanics" physics interface, and the energy equation uses the "Heat Transfer in Porous Media" physics interface. The computational model is set up as a two-dimensional axisymmetric physical model, with the center of the sample bottom as the coordinate origin, and the drainage surface at the $Z = 0.28$ m plane. The computational model and the locations of various measurement points are shown in Fig. 7.9.

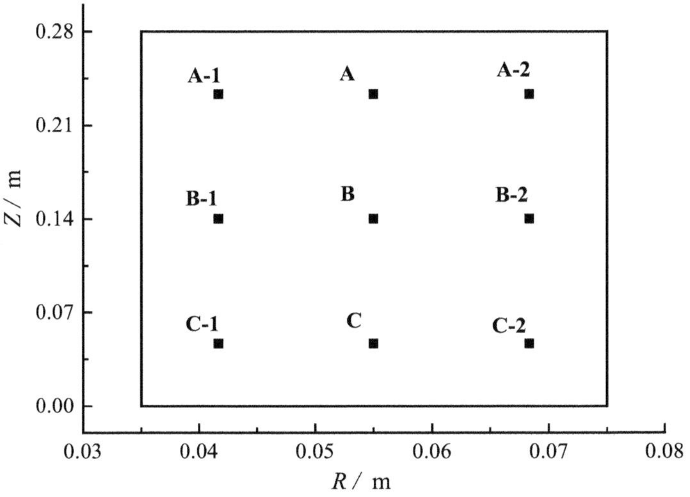

Fig. 7.9 Two-dimensional axisymmetric computational model

Table 7.2 Correction coefficient of Young's drained modulus

		$E_{b,0}$ (MPa)	Level 1 heating		Level 2 heating		Level 3 heating	
			η_T (1/K)	η_P	ξ	γ	ξ	γ
Path 1	$\sigma_3' = 50$ kPa	0.47	0.02	0.92	2.8	0.65	–	–
	$\sigma_3' = 100$ kPa	0.73	0.0145	0.775	2.2	0.65	–	–
	$\sigma_3' = 150$ kPa	0.51	0.0137	1.56	1.55	0.69	–	–
Path 2	$\sigma_3' = 50$ kPa	0.35	0.038	1.6	2.3	0.77	1.4	0.69
	$\sigma_3' = 100$ kPa	0.5	0.025	1.6	1.6	0.78	1.4	0.69
	$\sigma_3' = 150$ kPa	0.26	0.031	4.0	2.4	0.85	1.0	0.84

During the numerical calculation process, the changes in pore water pressure dissipation during each stage of the heating without drainage consolidation process and the overall average strain after isothermal drained consolidation are fitted. Table 7.2 presents the parameter values used in Eq. (7.78).

7.6.3 Pore Water Pressure Evolution Process

Figure 7.10 shows the evolution of the ratio of pore water pressure to confining pressure over time. According to the literature, the isothermal drained consolidation process in the experiment is completed quickly. For ease of comparison with the computational results, only the values from the first 20 min of the isothermal draining process are included in Fig. 7.10.

Figure 7.10 demonstrates that the patterns of pore water pressure dissipation in the experimental results are consistent with those in the computational results. It is observed that pore water pressure rises more rapidly during the initial stages of temperature transmission (i.e., when the temperature gradient is larger). The magnitude of pore water pressure during the stable phase of the heating consolidation

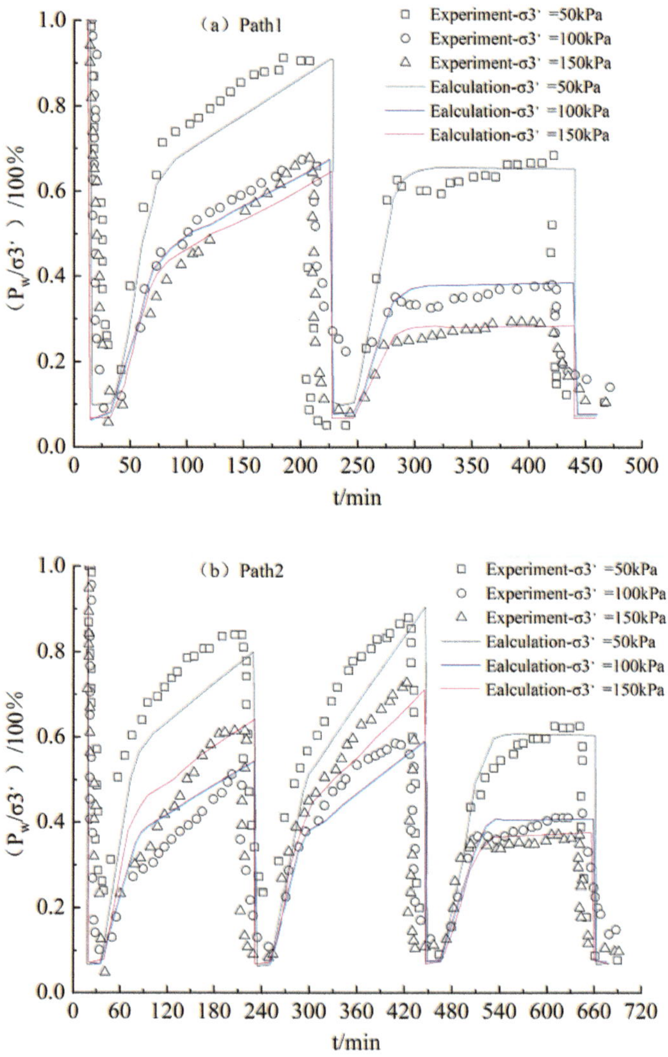

Fig. 7.10 Evolution of pore pressures with time

process is related to the size of the temperature gradient within the sample when temperature transmission is stable. Additionally, multi-stage heating consolidation can still generate relatively high pore water pressures. Earlier multi-stage heating processes have little impact on later stages of the heating consolidation process, such as the second level of heating in Path 1 and the third level of heating in Path 2. Since the relationship between temperature and time on the inner and outer surfaces of the hollow cylinder was simplified to linear in the calculations, and there are errors in parameter selection compared to actual values, there is a certain deviation between experimental and computational values.

The computational results indicate that the isothermal drained consolidation process is completed very quickly, generally stabilizing within 20 s. Figure 7.11 shows the evolution of pore water pressure over time during three drained consolidation processes under a confining pressure of $\sigma_3' = 100$ kPa in Path 1, as depicted in the dark gray area of Fig. 7.8a (measurement point locations as shown in Fig. 7.9).

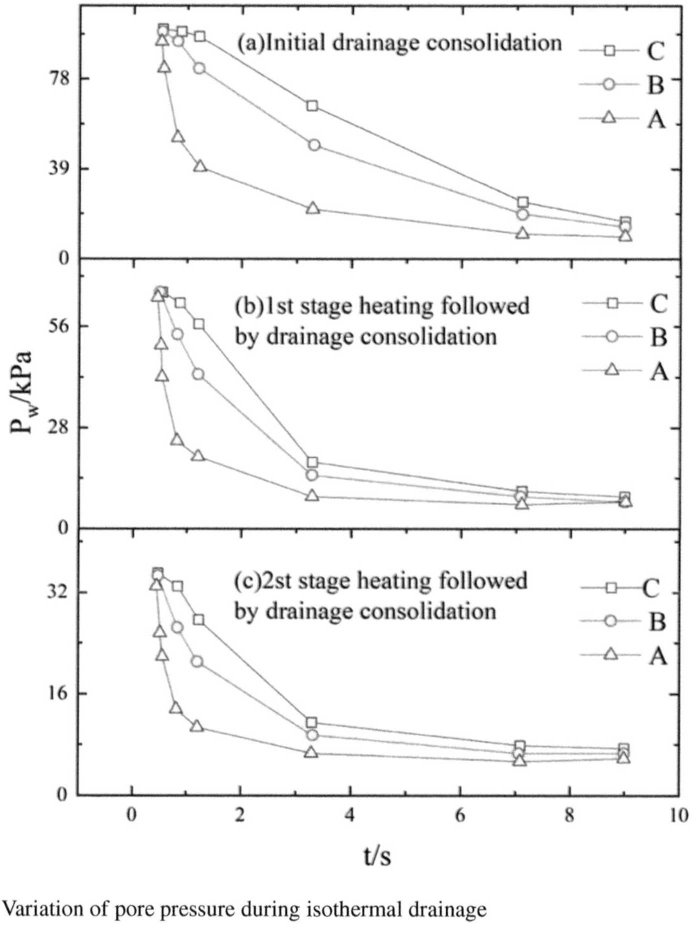

Fig. 7.11 Variation of pore pressure during isothermal drainage

Figure 7.11 illustrates that during the isothermal drained consolidation process, the dissipation of pore water pressure occurs rapidly, stabilizing by 9 s. Additionally, the closer to the drainage surface, the faster the water pressure dissipates. Since the pore water seepage equation accounts for the effect of gravity, the residual pore water pressure at the bottom of the sample is higher than at the top, with the bottom pore water pressure being approximately 2.5 kPa.

Figure 7.12 depicts a schematic diagram of the direction of pore water seepage velocity during the first level of heating at a confining pressure of $\sigma_3' = 150$ kPa in Path 1, as seen in the dark gray area of Fig. 7.8a. Figure 7.12 shows that after stabilization of the thermo-hydraulic action, pore water near the heat source in the soil sample flows from the bottom to the top, while in the outer soil body farther from the heat source, pore water flows from the top to the bottom, forming a recirculation under the influence of the top and bottom surfaces. The seepage velocity of pore water in the soil sample decreases gradually from the interior to the exterior, with average seepage velocities on the inner and outer surfaces being 7.97×10^{-7} m/s and 4.06×10^{-7} m/s, respectively. Moreover, in the Z-direction, the seepage velocity of pore water in the middle of the soil sample is greater than at both ends. Due to the presence of a 90° corner at the junction of the top surface, bottom surface, and side walls, the seepage velocity of pore water is minimized at this location, approximately 7.19×10^{-8} m/s. Similarly, in the R-direction, the seepage velocity in the middle is greater than at both ends.

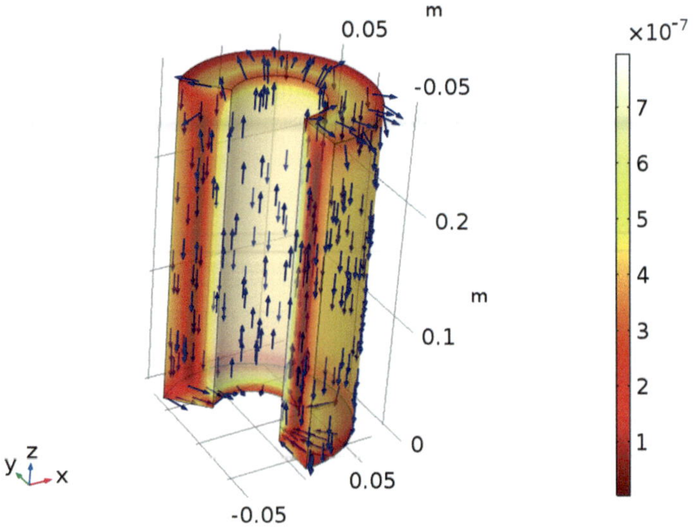

Fig. 7.12 Schematic of seepage velocity direction during heating and undrained consolidation

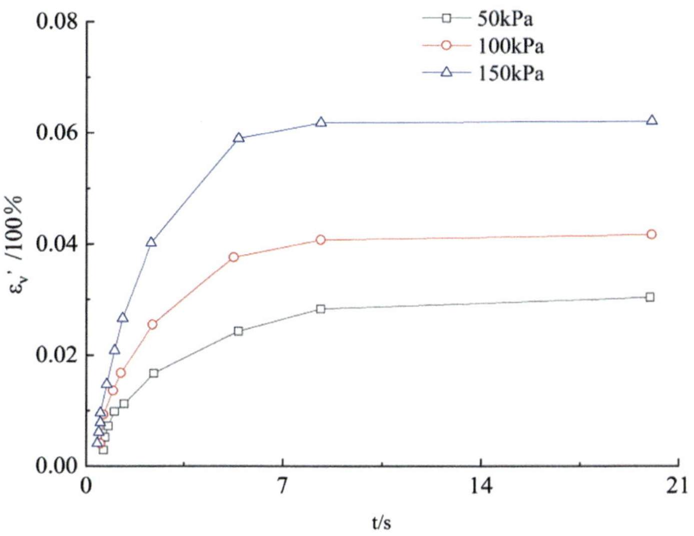

Fig. 7.13 Evolution of mean bulk strain with time at initial consolidation

7.6.4 Volumetric Strain Evolution Process

Figure 7.13 shows the trend of overall average volumetric strain $\varepsilon_{\mathrm{v}}'$ over time t during the initial drained consolidation under different confining pressures.

Figure 7.13 demonstrates that as the confining pressure increases, the average volumetric strain of the sample also increases. The volumetric strain stabilizes at values consistent with experimental results when $\sigma_3' = 50$ kPa, 100 kPa, and 150 kPa, with initial consolidation strains of $\varepsilon_v' = 0.0292$, 0.0405, and 0.0612, respectively. Alongside Fig. 7.11, it is evident that the change in volumetric strain is significantly influenced by the dissipation of pore water pressure. The rate of change in volumetric strain is initially high and gradually stabilizes.

Figure 7.14 shows the evolution of volumetric strain ε_{v} during different drained consolidation processes under a confining pressure of $\sigma_3' = 100$ kPa in Path 2.

In Fig. 7.14, it is observed that during the drained consolidation process, volume contraction first occurs at the top part of the soil sample, where the top surface is the drainage face, due to the dissipation of pore water pressure. As the consolidation time progresses, the region undergoing volumetric change gradually extends downward until the consolidation process is complete. Additionally, Fig. 7.14c indicates that after the isothermal drained consolidation process is complete, the internal volumetric strain of the soil sample is slightly less than the external. The internal surface volumetric strain is approximately $\varepsilon_{\mathrm{v}} = 0.0398$, while the external volumetric strain is about $\varepsilon_{\mathrm{v}} = 0.0403$. Figure 7.14d shows that after the first heating consolidation, the volumetric strain is not much different from that after the initial drained consolidation. When drainage consolidation occurs again, deformation first takes place at

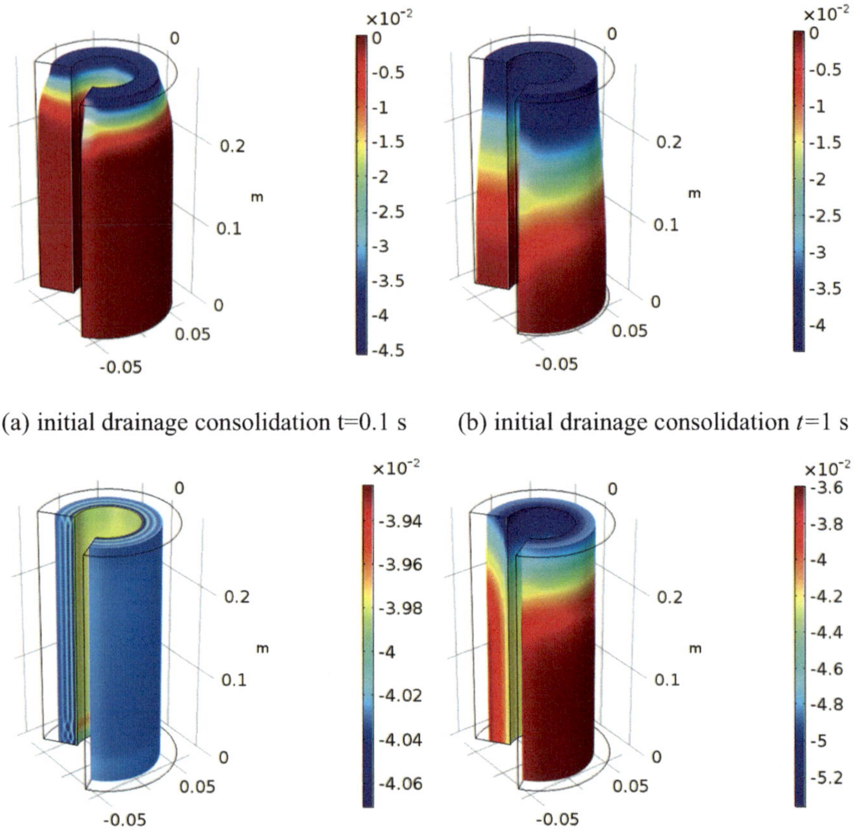

(a) initial drainage consolidation t=0.1 s (b) initial drainage consolidation t=1 s

(c) initial drainage consolidation t=1200 s (d) initial drainage consolidation t=0.1 s

Fig. 7.14 The evolution law of average body strain with time in initial consolidation

the top of the soil sample and then extends to the bottom. Furthermore, due to the accumulation of volumetric strain from the previous two phases, the range of volumetric strain change during this process is smaller than during the initial drained consolidation.

The dissipation pattern of pore water pressure at average volumetric strain Z and point A during the initial consolidation process, when the permeability K in the porous medium is set to K_0, $0.5K_0$, $0.1K_0$, and $0.05K_0$ ($K_0 = 6.48 \times 10^{-12}$) while other parameters remain unchanged (under Path 1 with $\sigma_3' = 100$ kPa), is shown in Fig. 7.15.

Figure 7.15 illustrates that as the permeability K decreases, the dissipation of pore water pressure during the drained consolidation process becomes slower, meaning that water penetrates the porous medium more slowly. This observation is consistent with the basic principles of soil mechanics related to seepage. Furthermore, the rate of change in volumetric strain decreases as K decreases. However, changes in

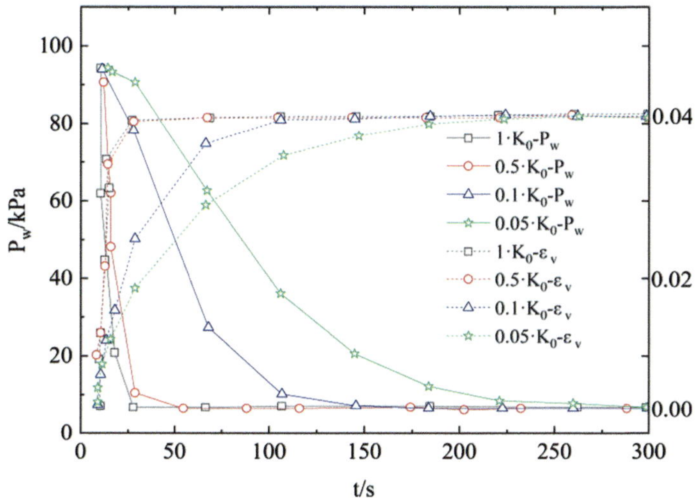

Fig. 7.15 Evolution of $\varepsilon_V{}'$ and P_W with K at initial consolidation

permeability do not affect the volumetric strain values when consolidation stabilizes. This implies that the rate of pore water seepage in the porous medium does not influence the final volumetric strain of consolidation, but merely alters the rate of change in volumetric strain. After the second drained consolidation, the internal volumetric strain is less than the external volumetric strain.

After the initial consolidation, the evolution of average volumetric strain (in this case, the volumetric strain neglecting the impact of the initial consolidation strain) under two heating paths is shown in Fig. 7.16. When the confining pressure is higher, the average volumetric strain during thermal consolidation is smaller (because the initial consolidation strain is larger under higher confining pressure, thus the actual volumetric strain is greater). This indicates that higher confining pressures tend to inhibit further deformation of the solid matrix during the thermal consolidation process. The stages with significant changes in volumetric strain in Fig. 7.16 occur during the drained consolidation process, while during the heating consolidation process, the volumetric strain tends to decrease (i.e., volumetric expansion occurs). However, this trend is weak. Due to the thermal expansion effect of the solid matrix, volumetric strain decreases. An increase in pore water pressure during the heating consolidation process compresses the solid matrix, leading to an increase in volumetric strain. The interplay of temperature effects and pore water pressure effects results in minor changes in volumetric strain during the heating consolidation process. Figure 7.16 shows that under Path 2 with confining pressures $\sigma_3' = 50$ kPa, 100 kPa, and 150 kPa, the volumetric strains caused by the first level of heating are $\varepsilon_v' = 0.011$, 0.009, and 0.008, respectively, and for the second level of heating, $\varepsilon_v' = 0.026$, 0.022, and 0.020, respectively. These computational results align with the experimental results, which were $\varepsilon_v' = 0.0112$, 0.0092, and 0.0072, and $\varepsilon_v' = 0.026$, 0.0223, and 0.0181, respectively. Due to the rapidity of the isothermal drained consolidation process, the average volumetric strain in Fig. 7.16 appears to rise sharply.

Under Path 2, with a confining pressure $\sigma_3' = 50$ kPa, the evolution of the volumetric strain ε_v at each measurement point is shown in Fig. 7.17.

Figure 7.17 shows that there are significant differences in the trend of volumetric strain changes across the same cross-section (i.e., coordinates in the Z-direction, represented by the same color in Fig. 7.17). During the first level of heating, the volumetric strain on the inside of the sample increases (i.e., the matrix is compressed). This is due to the interior side experiencing a faster and higher temperature increase compared to the exterior, which reduces the drained bulk modulus E_b due to the greater influence of temperature, thus increasing the matrix volumetric strain under

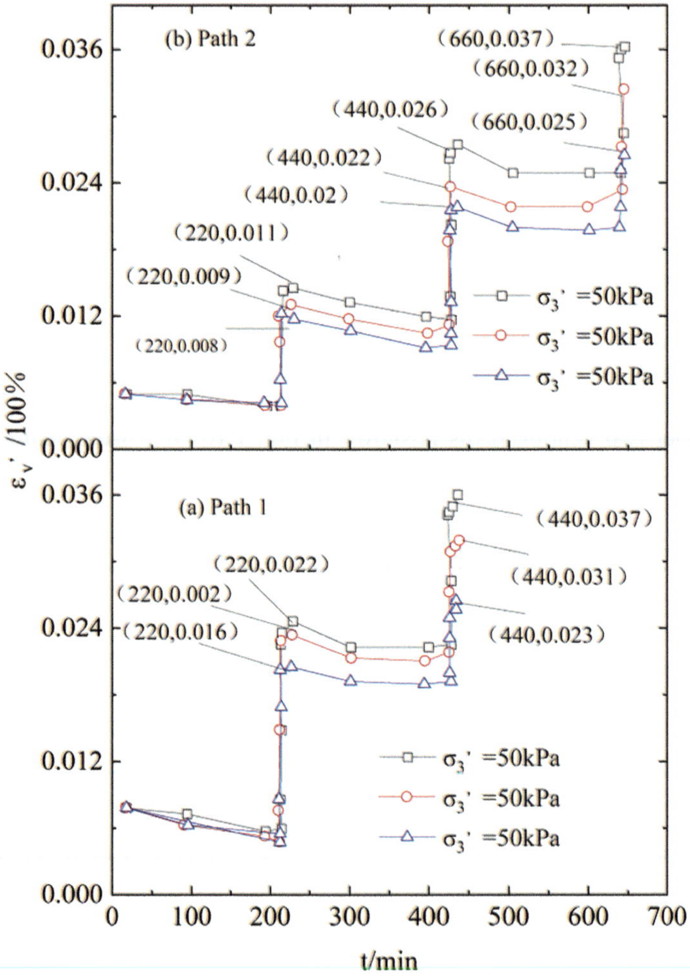

Fig. 7.16 Evolution of mean bulk strain with time

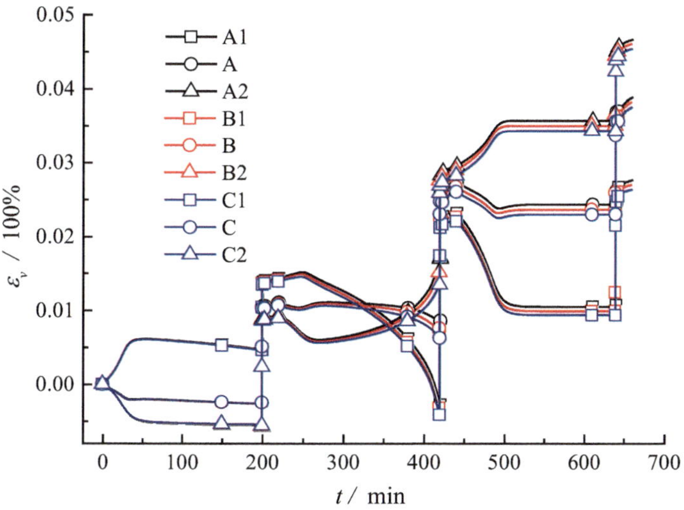

Fig. 7.17 Evolution of bulk strain at measured points with time

pressure. However, the exterior side of the sample experiences a slower temperature increase and smaller gradient, and the water pressure increases relatively faster, leading to a weaker trend (or even an increase) in the reduction of E_b. Additionally, the water pressure offsets some of the confining pressure, resulting in a decrease in the exterior volumetric strain (i.e., the matrix undergoes expansion). After the first drained consolidation, the interior volumetric strain is greater than the exterior strain.

During the second level of heating, the change in volumetric strain in the middle position (across the cross-section) of the sample is minor, showing a decreasing trend. The changes in the interior and exterior volumetric strains are more significant and opposite to the trends observed during the first level of heating. The interior volumetric strain of the solid matrix initially increases and then decreases, while the exterior strain first decreases and then increases, ultimately exhibiting expansion on the interior and compression on the exterior. After the second drained consolidation, the exterior volumetric strain is greater than the interior strain.

In the third level of heating, expansion still occurs near the interior side of the sample, while the exterior volumetric strain continues to increase. After the third drained consolidation, the exterior volumetric strain is greater than the interior strain, and the difference is larger than the results after the second drained consolidation.

Additionally, Fig. 7.17 also indicates that the trend of volumetric strain changes along the same longitudinal section (i.e., coordinates in the R-direction) is consistent. Throughout the process, the volumetric strain at the lower part of the sample is generally slightly less than at the upper part. This is due to the effect of gravity. Under the influence of gravity, the pore water pressure at the lower part of the sample is higher than at the upper part, meaning that the lower part offsets more of the confining pressure (i.e., the actual force acting on the solid matrix is smaller).

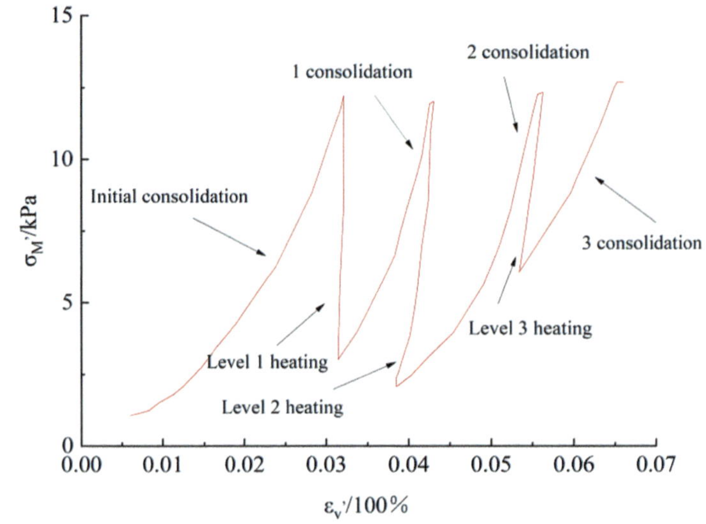

Fig. 7.18 Evolution of mean miss stress with mean bulk strain

Therefore, the lower part's volumetric strain is slightly less than that of the upper part.

Figure 7.18 presents the evolution of the average volumetric strain ε_v' and average Miss stress σ_M' under Path 2, with a confining pressure $\sigma_3' = 50$ kPa.

Figure 7.18 shows that the changes in σ_M' relative to ε_v' during the drained consolidation process are more gradual compared to the heating consolidation process. This is because during the drained consolidation process, the dissipation of pore water pressure in the soil pores leads to rapid changes in volumetric strain. In contrast, the average volumetric strain during the heating consolidation process is smaller.

During the drained consolidation process, as the average volumetric strain increases, the slope in Fig. 7.18 also increases, indicating that the elastic modulus of the compacted soil gradually increases. Furthermore, as the temperature load levels increase, the average volumetric strain progressively increases, consistent with experimental conclusions. This reflects the compressive effects and consolidation behavior as the soil adjusts to both mechanical and thermal stresses, validating the interplay between the soil's mechanical properties and the environmental conditions imposed during the tests.

7.7 Summary

This chapter derives a theoretical model for the coupling of temperature, seepage, and deformation fields, which takes into account variations in medium density and porosity with temperature and pressure. The model employs a nonlinear elastic constitutive equation and incorporates heat source terms such as viscous dissipation. Using the provided governing equations for thermo-consolidation of saturated soils and the state equations, the thermo-hydro-mechanical coupling response characteristics during thermo-consolidation tests are analyzed. The proposed model effectively explains the dissipation process of pore water pressure and the volumetric strain response characteristics observed in thermo-consolidation experiments.

Chapter 8
Conclusion

A self-developed experimental device with controllable temperature path and load conditions was developed to address the thermal solidification problem of saturated clay. Based on the experimental device, the thermal solidification characteristics of saturated clay under different load and temperature path conditions were studied, including the evolution laws of pore water pressure and thermal deformation. At the same time, based on the theories of thermoelasticity, thermal solidification constitutive models for saturated clay were established and their effectiveness was verified. The specific conclusion is as follows:

(1) Two different thermal bonding test devices have been developed, which are suitable for cylindrical specimens and hollow cylindrical specimens, respectively. They can be used to measure the thermal bonding deformation and pore water pressure changes of specimens under the coupling effect of changing temperature loads and external loads. The temperature inside the pressure chamber of the experimental device can vary within the range of 17–100 °C and can also withstand high isotropic surrounding consolidation pressure (0–3 MPa). The test apparatus for hollow cylindrical specimens can independently control the temperatures of the inner and outer boundaries of the specimen, maintain a radial temperature gradient within the specimen, and apply radial consolidation pressure. It is suitable for conducting tests under specific coupling paths of temperature and stress.

(2) The changes in pore water pressure and strain of the saturated specimen under different effective consolidation pressures are similar. During the process of temperature increase, the pore water pressure caused by temperature continuously increases. When the temperature reaches the set temperature, the pore water pressure inside the sample also reaches its peak, and with the increase of effective consolidation pressure, the peak pore water pressure seems to slightly increase. In addition, the consolidation drainage after temperature increase is much greater than the water absorption after temperature decrease, that is, after

G. Yang and B. Bai, *Experimental and Theoretical Study on Thermal Consolidation Behavior of Saturated Clay*, https://doi.org/10.1007/978-981-96-6909-7_8

the process of heating and cooling, the sample undergoes significant shrinkage deformation overall.

(3) During the multi-stage heating and subsequent cooling process, the solidification drainage caused by the dissipation of pore water pressure induced by temperature is greater than the water absorption at the initial temperature, ultimately resulting in irreversible volume shrinkage deformation. On the other hand, under the conditions of no drainage between multiple heating stages and gradual cooling, the pore water pressure after temperature increase can even reach the confining pressure value, and when it drops to the initial ambient temperature, the sample still maintains a large pore water pressure, which is also an irreversible thermodynamic process.

(4) For the thermodynamic response of saturated silty clay caused by repeated heating and cooling processes under undrained conditions, research has shown that the temperature change process of the sample is consistent in each cycle throughout the entire heating and cooling process of saturated silty clay, regardless of confining pressure and the number of temperature cycles. During the temperature rise process of each cycle, the pore water pressure continuously increases and tends to a stable value, and the growth trend is similar. Meanwhile, the normalized maximum pore water pressure increases with the increase of the number of cycles. During the cooling process, as the number of temperature cycles increases, the rate of decrease in normalized pore water pressure seems to slow down, while the residual pore water pressure gradually increases. In addition, as the confining pressure increases, the normalized residual pore water pressure tends to decrease. As the temperature increases, the thermal expansion effect of pore water and solid particles leads to a significant increase in the volume of the sample, while as the temperature decreases, the volume of the sample continuously decreases and approaches zero.

(5) The experiment on the thermal solidification characteristics of saturated cohesive soil during the process of repeated hot cold changes shows that as the sample temperature increases, the rate of temperature rise gradually slows down, while the opposite is true during the cooling process. Moreover, the temperature change process is independent of the number of temperature load cycles and the confining pressure exerted on the specimen. During the process of non-draining heating, significant pore water pressure is induced, and the pore water pressure induced by the second temperature cycle is smaller than that induced by the first temperature cycle. During the process of temperature drop without drainage, negative pore water pressure is first generated in the sample. When the temperature drops to the set temperature value, the pore water pressure begins to rise again and then converts into positive pore water pressure. If the sample is allowed to drain (or absorb water), the isothermal process of the sample after heating and cooling generally exhibits a shrinkage drainage process, and the drainage strain caused by heating is much greater than that caused by cooling.

(6) Regarding the thermal solidification characteristics of isothermal single-stage and multi-stage heating cooling paths at the inner and outer boundaries of the sample, as well as non-isothermal multi-stage heating cooling paths at the inner

and outer boundaries, experiments have shown that different heating levels and heating sequences result in significant differences in the maximum value of pore water pressure after stabilization under the same final temperature load, as well as different body strains. In fact, this reflects the significant influence of the temperature load path on the thermal solidification effect of the sample.

(7) The experimental study on the influence of overconsolidation ratio on the thermal solidification characteristics of hollow cylindrical saturated silty clay under different temperature paths shows that the maximum pore pressure generated by samples with different overconsolidation ratios during the same heating process is basically the same. However, during the same cooling process, with the increase of OCR, the absolute value of the negative pore pressure generated tends to increase. The volumetric strain generated during the heating or cooling process decreases exponentially with the increase of OCR, and the larger the temperature amplitude, the greater the magnitude of this change. The final volumetric strain after the heating cooling process also decreases with the increase of OCR; The more temperature levels set during the process of heating or cooling to the same temperature value, the greater the strain of the solidified body. Moreover, as OCR increases, the difference in strain between different levels decreases, while OCR has a relatively small impact on the difference in final strain between different levels of heating and cooling.

(8) Based on the theory of thermoelasticity, a thermal solidification model was developed that considers the changes in medium density and porosity with temperature and pressure and adopts a nonlinear elastic constitutive equation. The established nonlinear thermal bonding model is in good agreement with the experimental results. The model can effectively explain the dissipation process of pore water pressure and the characteristics of volumetric strain response in thermal solidification tests, indicating that the change in pore water permeability in saturated soil samples has a significant impact on the evolution process of water pressure dissipation and volumetric strain.

GPSR Compliance

The European Union's (EU) General Product Safety Regulation (GPSR) is a set of rules that requires consumer products to be safe and our obligations to ensure this.

If you have any concerns about our products, you can contact us on ProductSafety@springernature.com

In case Publisher is established outside the EU, the EU authorized representative is:

Springer Nature Customer Service Center GmbH
Europaplatz 3
69115 Heidelberg, Germany

Batch number: 10704307

Printed by Printforce, the Netherlands